George B. Starkweather

The Law of Sex

being an exposition of the natural law by which the sex of offspring is controlled in

man and the lower animals

George B. Starkweather

The Law of Sex
being an exposition of the natural law by which the sex of offspring is controlled in man and the lower animals

ISBN/EAN: 9783337231378

Printed in Europe, USA, Canada, Australia, Japan

Cover: Foto ©berggeist007 / pixelio.de

More available books at **www.hansebooks.com**

THE

LAW OF SEX:

BEING

*AN EXPOSITION OF THE NATURAL LAW BY WHICH
THE SEX OF OFFSPRING IS CONTROLLED
IN MAN AND THE LOWER ANIMALS.*

AND GIVING THE

SOLUTION OF VARIOUS SOCIAL PROBLEMS.

WITH FORTY ILLUSTRATIVE PORTRAITS.

BY

GEORGE B. STARKWEATHER, F.R.G.S.

LONDON:

J. & A. CHURCHILL,

11, NEW BURLINGTON STREET.

—

1883.

PREFACE.

IN the following pages I claim to make known a
new discovery of a great law of Nature; nothing
less than the law which governs the sexes, and
whereby the sex of offspring can be controlled.

A satisfactory knowledge of the law of sex
would largely add to the happiness of many families
in all civilized countries, and also help us to the
solution of many intricate social problems of urgent
character, such, for instance, as the possible assimilation of alien and lower races, the avoidance of a
lamentable redundancy of women in old settled
countries, etc.; and I am sanguine enough to believe
that the accomplishment of some of these results will
be brought about by the publication of this volume.

I have spent twenty years of my life in studying
the subject and its connected literature in various
languages, and though my theory has been matured
for several years, I have delayed its publication till

I could, by extensive observation, establish its truth beyond the possibility of doubt.

Why do some families consist entirely of boys, and others of girls? We see many such instances, and they must have a meaning: they are causal, not casual. Sex and reproduction are themes almost inseparably allied, the gravity of which cannot be over-estimated, as affecting the future of the race. Yet, though the value of *Education* is practically felt, *Generation*, the foundation on which the whole superstructure of education must inevitably rest, has been regarded as too vague or vulgar in its nature to merit consideration. How long is such a state of things to continue? In my opinion, only until a rational hypothesis, fully elucidating the question of offspring, has been given to the world. In the present work, which is original throughout, I maintain that this desirable end is accomplished.

In considering a subject of this intricate nature, mere assertions will obviously carry but little weight: I have therefore been careful to fortify every position with ample proof, both from the highest scientific writers and from the realm of facts. Many scores of authorities are quoted; still more numerous examples are cited. My theory being based upon general physiological laws, I have been able to avoid all objectionable detail.

The work does not claim to be perfect in style, though I am confident that my arguments and deductions will prove conclusive, and will not fail to carry conviction. The *facts*, too, will be intelligible to every reader. It is therefore hoped that any defects in arrangement, or faults of diction, will be overlooked in consideration of the important truths set forth.

G. B. STARKWEATHER.

London, July 1883.

CONTENTS.

CHAPTER I.

INTRODUCTORY.

CHAPTER II.

THE PROBLEM STATED.

CHAPTER III.

FALLACIES OF CURRENT THEORIES.

CHAPTER IV.

THE NATURE OF SEX AND SEXUAL EQUALITY.

CHAPTER V.

HEREDITY AND STERILITY.

CHAPTER VI.

"SUPERIORITY" THE CONTROLLING PRINCIPLE OF SEX.

CHAPTER VII.

WHAT INDICATES AND DETERMINES " SUPERIORITY."

CHAPTER VIII.

VERIFICATION FROM OBSERVED FACTS.

CHAPTER IX.

A UNIVERSAL BALANCE OF SEX THE PROOF OF MY THEORY.

CHAPTER X.

PRACTICAL RESULTS—CONTROL OF SEX.

CHAPTER XI.

OTHER SOCIAL PROBLEMS—ELEVATION OF THE RACE—CONCLUSION.

STIRPICULTURE.

CHAPTER I.

INTRODUCTORY.

AMONG all the many questions connected with heredity and the perpetuation of our race upon which light has been thrown by modern scientific research, there is probably not one which more deeply interests each individual than the question of the sex of his or her offspring. And on a par with this question is that of the probable dispositions of those who are to continue our name, and make or mar the happiness of our parental life. Certainly no question of social science has hitherto been involved in such complete darkness as that of the law of sex. Just in proportion to the keenness of our desire to learn something definite upon the subject, has been the completeness of the obscurity which baffled us.

There has, however, been no lack of effort to penetrate the mystery, and from the earliest times to our own, and in all parts of the world, every age has found some kind of answer to the riddle, as piety or credulity, acuteness or cupidity, prompted.

On no subject has charlatanism had more to say, nor is there any upon which its declarations have met with so favourable a reception; no scientific question has attracted the thoughts of greater men; and yet the theories advanced are as diverse as they are numerous, and as contradictory as they are improbable. They have but one feature in common, that is, their utter inability to account for the facts.

Burdach, the distinguished German physiologist, has done posterity a service by compiling a list of more than a thousand writers who have studied this question during the past twenty-five centuries, nearly all of whom have framed theories of their own, or have supported some of those already enunciated, and all of whom have passed away without suggesting or discovering anything to disperse or even to diminish the darkness in which the whole subject was enveloped.

Ever since the days of Aristotle, Plato, Socrates, Hippocrates, and Galen, it has been a favourite theme of discussion and speculation with the greatest minds the world has produced—among philosophers no less than physicians. Buffon, Haller, Bonnet, Cuvier, Priestley, Lamarck, Carus, Oken, Wolff, Blumenbach, and Von Baer, are among the more recent names, together with Darwin and Spencer of our own times.

From the very dawn of Natural Science down to the present time, most of her votaries are known to have given attention to the solution of this interesting, yet, as it seemed, almost hopelessly recondite law of Nature.

After such an expenditure of time and thought, it is humiliating to confess that, on the whole, the question occupies a position infinitely worse than it did when Aristotle first considered it; for, so far from retaining the place which its undeniable value and dignity demand, it has fallen into the hands of the mountebank, who therewith imposes upon the credulity of the ignorant and superstitious. To such an extent is this the case that few are willing to have their names identified with it, and the *American Agriculturist*, one of the most reliable and respectable publications of its class, has declared that " whoever pretends that he can determine sex at will is either a fool or a knave " ; and the professional astrologer stands, probably, higher than he in popular esteem to-day.

So absurd are most of the explanations hitherto offered that society has virtually tabooed the subject, until some authoritative theory shall claim attention by a confident appeal to facts; and even modern science, in spite of the amazing boldness with which it has attacked, and often solved, many knotty questions in Nature, has had as yet little or nothing to say on this, perhaps the most important of all.

In the *Descent of Man*, Dr. Darwin himself, who may for the nonce be taken as the spokesman of modern science, not only alludes to "the unknown causes which determine the sex of the embryo," but concludes his summary of evidence as to the proportion of the sexes and the causes controlling their production, with the following confession of the inability of science to solve the enigma:—

"I formerly thought that when a tendency to produce the two sexes in equal numbers was advantageous to the species, it would follow from natural selection, but I now see that the whole problem is so intricate that it is safer to leave its solution for the future."*

From this it is safe to assume that hitherto no positive knowledge has been acquired on this vital question; but where scientific knowledge is wanting, empirical views are sure to abound, and consequently, as might be expected, numerous plausible theories have been promulgated, which have been to a great extent accepted by the unthinking as they successively appeared; but those more deeply interested, after testing different theories and proving their fallacy, generally give up the problem, and perhaps conclude that sex is ruled by the arbitrary decision of an overruling and inscrutable power—a decision fatal

* "Descent of Man," c. viii., Appendix.

to the further investigation of the subject. Why has the obscurity surrounding this subject prevailed so persistently? Why has it not received a share of that illumination which has been thrown into some of the remotest corners of the field of Nature? I believe that an answer may easily be given.

In olden times the prevalence of the deductive—*a priori*—method of research barred all progress in real knowledge of the laws of nature and those of life. It was difficult and tedious to observe and record, and easy to meditate and theorize; consequently after eighteen hundred years the science of Bacon's day was not much in advance of that of Aristotle: but Bacon's advocacy of the inductive—*a posteriori*—method, led men to study Nature by observation and experiment, and progress became at once marked and continuous.

Yet in spite of the great strides in knowledge during the last two hundred years, in the matter of the predetermination of sex no advance, or but a very slight one, has been made; and although since the publication of Von Baer's great work, embryology has become an independent and important science, and all the processes of the development of the embryo have been accurately studied, yet no light whatever, of a reliable character, has been thrown upon the causes determining its sex, and only a few scattered hints, such as Darwin's theory of Pan-genesis, have been suggested as to the operation of the causes controlling the dispositions and constitutions of children.

As a subject for special investigation, the one before us—though not without its difficulties and discouragements—possesses some inviting features. Unlike most other sciences, it is ever accessible to the student. Astronomers are always confined to their observatories, telescopes, etc., and must even go to the South Sea Islands to watch the transit of Venus. And in the same manner botanists, entomologists, zoologists, indeed those who are devoted to any department of Natural Science, are obliged to set out in search of virgin fields for specimens. But in seeking to discover the law of sex we find material to work upon everywhere; in every social gathering. in every household and in every library may be found data with which to construct a hypothesis. The superabundance of facts—and those sometimes of a most conflicting nature—fairly bewilders one, and constitutes the chief impediment to success. A skilful detective, with no other clue than the fragment of a letter, a faded picture, or a lock of hair, will spend months in working up his case, and at last bring it to a successful issue. Should our problem, then, be abandoned because of an over-supply of evidence? Surely not. The abundance of evidence and the ease with which our theories can be brought to the test of fact ought rather to inspire us with a confidence that the truth must be discoverable, and to spur us on to a systematic collection and collation of the evidence so readily available.

We may draw encouragement from the fact that time is on our side. The centuries tell upon the subject before us as upon everything else.

> " Like some old castle, battered and decayed,
>   New light strikes in through chinks which Time has made."

We owe much to the age in which we live, as well as to the past; and slowly but surely the Time-spirit, as the Germans say, will whisper the secret into our ears if we will but listen. On this subject science is listening and scrutinizing earnestly enough at the present time, though hitherto not quite in the right direction. We live in an age of observation and careful study of Nature; she will not fail us in the end.

We cannot force Nature, but with a knowledge of her laws we can lead her.

"We conquer Nature by obeying her."

A knowledge of this subject is to be gained only, as it has been gained on all other subjects, by strict induction, by observation, and verification.

The whole result of modern scientific research and discovery is to impress us with still increasing conviction, that the phenomena of the universe, including those of life, are regulated and immutable, depending upon relations of cause and effect; and subject, as the Duke of Argyll has said, to a majestic and universal " Reign of Law," the principles of which are discoverable by us if we only apply the requisite patience and enlightened study; and this is now being brought to bear upon almost every province of the domain of Nature.

Accepting this conclusion, it would seem that there is not sufficient reason for believing, as many do, that the prepotencies capable of determining sex pertain to the realm of the unknowable—precisely the point I wish to press here.

In most existing theories the known inequalities

between the sexes are assumed to be natural and unavoidable; that is to say, we throw upon Nature the responsibility of the evils consequent upon these deviations from equality; whereas I hope to show that she is here, as everywhere, simply but sternly just, pardoning nothing that is done amiss, but granting no favours, and inflicting no gratuitous penalties. I maintain that the evils which arise from the inequalities of sex, nay, the very inequalities themselves, are of our own creation, and as we have transgressed Nature's laws so we have to pay the penalty she exacts. Nay, further, I assert that not only is the preponderance of one sex over the other preventible in society at large, but that it is perfectly possible to insure the sex of our offspring with something approaching to absolute certainty. This is a bold assertion, and one which, if established, would be productive of immense good, national and individual.

What would not the royal, noble, and aristocratic families of the primogenital order give for some secret by which they could with some degree of certainty ensure that their firstborn should be of the sex desired? Such knowledge would give all parents their choice, and thus conduce greatly to human happiness.

I confidently believe that those who will bear with me to the end will confess, on closing this volume, that my assertion is not presumptuous; that the theory here advanced, though it may yet lack absolute precision in its details, is certainly in accordance with the evidence—as each may ascertain for himself, by observation in his own immediate circle. Besides explaining the most apparently contradictory facts, my theory renders prediction feasible,—and that is the surest proof of the truth of any hypothesis; and more than this, it fully explains the maintenance of an approximate balance of sex throughout the world, which no previous theory has been able to do; and to this aspect of the subject I shall devote an entire chapter.

CHAPTER II.

THE PROBLEM STATED.

What law determines sex in general—General opinion of the scientific world on the question—Reliable records hitherto confined to Europe and America—Statistical Evidence cited from Darwin's *Descent of Man*—Nature credited with cunning devices to equalise the sexes—Ingenious theories to explain Nature's diversity of operation—A "curious natural law of sex"—Excess of male births estimated at 4 per cent.—Excess of female births during the cholera epidemic in Philadelphia—Similar fact observed in Paris—Great excess of male births at Buenos Ayres after cholera plague—Yet nature strangely accredited with the readjustment—The "natural tendency" to an excess of male births shown to be a fallacy—Excess in mortality of male infants—Erroneous explanations—Numerical relation of males to females at various ages—Polygamy does not affect the proportion of the sexes—Darwin's assumption that the law of sex is universal—The rational theory of the law of sex must rest on a broad basis—Striking difference between different races—Excess of male births among the Jews—Tendency to equalization of sexes in our colonies—Tendency of Nature towards equality of sexes at birth—Absurd processes of adjustment accredited to Nature—Variation of percentages of male and female births in town and country—True indications of plan of adjustment—Tendency of Nature to restore a disturbed equilibrium of the sexes—Various secondary causes produce local inequalities—The theory of the more love determining the sex a harmless assumption—Sex determined at the time of conception—Experiments with caterpillars and moths—The true theory must apply to men and animals—Facts apparently inharmonious will be found to support the true theory of sex.

ALTHOUGH the avowed object of this work is to discuss individuality of sex,—that is, the causes controlling the sex in each family,—yet the consideration of this subject cannot be isolated from the larger question of which it forms a part.

The larger problem, briefly stated, is "What law or laws determine, directly, or by their mutual interaction, the production of the sexes in general? Now if it could be shown that there is an overruling and inscrutable law which governs the total production of either sex in each country or throughout the world irrespective of local and personal causes—which according to our view might seem likely to affect the proportional numbers,—it is scarcely probable that the smaller question of the determination of the proportion of the sexes in a single family would be independent of this law. If the sex of offspring were thus settled beforehand, it would be futile to talk of controlling the sex, and therefore if we hold that the determination of sex is a matter raised by such a higher law, above the operation of those local factors over which we have any control, we must of course relinquish our investigation of the subject.

That I do not hold this opinion is evident from the publication of this book, and yet there seems to be a tacit assent given by the scientific world to the opinion that some such higher law does obtain. There seems to be, it is said, " a natural tendency " to a preponderance of males at birth, although no explanation of this tendency is attempted; it is in fact received as an ultimate axiom which cannot be questioned. This is, however, only another way of saying that there is a preponderating effect of which we do not know the cause. Just in the same way, before the discovery of the law of gravitation men spoke of the " natural tendency of all bodies to fall to the ground."

We will now consider the evidence relied upon to support the above conclusion, that the proportion of the sexes is determined by some innate tendency as fixed and irresistible as "the tendency of small bodies to move towards great ones." Such evidence is for the most part statistical, and as reliable records have hitherto been confined to Europe and America, and a few isolated districts elsewhere, it is mainly on these that the opinion in question is founded. The nations, therefore, which exhibit the phenomenon are all in approximately the same social condition, all live under the same kind of laws and customs, and under nearly the same conditions of climate. The evidence given by Darwin (*Descent of Man*, c. viii, Appendix) may be cited as an example, from which the following facts are extracted :—

England during ten years showed an excess of males over females at birth, represented by the proportion of 104·5 males to 100 females. France for forty-four years shows 106·2 to 100; Russia has a still higher percentage of males, 108·9 to 100; Philadelphia, U.S.A., shows a 10 per cent. excess,— 110·5 to 100; and the Jews 120 to 100; while among the Christian populations of the same districts the proportion is only 104 to 100.

From computations of this kind, of which there is abundance in every statistical almanac, it is generally assumed that nature produces the sexes in unequal proportions; and this is repeated until probably every reader has frequently heard that children are born in a proportion not varying greatly from *twenty-one* males to *twenty* females, the surplus boys being wisely provided by Nature to meet the deficiency arising from accidents and violence to which males are chiefly exposed. This idea is found in so many scientific works, and recognized as accurate by such high authorities, that few would hesitate to accept it as a fundamental truth.

Then let us note the wonderful way in which the theorizers accredit nature with all sorts of righteous and cunning devices, so as to account for this unequal production of the sexes of which they suppose they have convicted her.

Mr. Henry Beverley, M.A., Inspector-General of Registration in Bengal, writes to the *Journal of the Statistical Society*, that, "it has long been recognized as a law of Nature that while slightly more males are born than females,—say an excess of 5 per cent.,—in the sexes of adult age we find nearly a corresponding excess of females." He quotes the census returns of the several European powers which establish this fact. In India—in the Punjab and other North-West Provinces—on the contrary, he shows that adult males predominate, but that it is not absolutely certain whether they are born in excess or not.

Several ingenious theories have been put forth to explain this contradiction between the operations of Nature in the East and West. Mr. Plowden, Fellow of the same Society, in his official report, contends that a warm temperature increases the male preponderance in proportion as we approach the tropics. Mr. Beverley, however, doubts the assumption that more males are born in India, as no reliable birth record is kept in that country.

Mr. D. A. Wells, well known in America for his interest in social science, in his work entitled "Things not Generally Known," says: "There is a natural law of relation between the sexes, which is found to vary at different ages according to the different dangers to which they are exposed. This is one of the most curious of natural laws, and one of the most interesting—demonstrating the admirable economy of adaptations between the several parts of the natural system. If the numbers of males and females born were exactly equal, the result would be, that before they reached middle age the male sex would be reduced too low, and become inadequate to the purposes which it has to fulfil. In fact, the number of males born is always greater than that of females by about 4 per cent."

The *American Journal of Medical Science*, for July 1848, states that Dr. Emerson, of Philadelphia, about that time chanced to discover that, during the cholera epidemic of 1832 in his city, instead of the usual surplus of male conceptions, there was quite an excess of the opposite. It also appears that a similar anomaly was noticed in Paris during the epidemic of the same year; and what perhaps tends to make the circumstance more remarkable is, that this deviation was strictly confined to those portions of the two cities which were visited by cholera. These facts were proclaimed far and near as new evidences of that wonderfully intelligent and recuperative faculty of Nature by which she readjusts her forces, and preserves a constant equilibrium.

The following figures, as will be seen, do not accord very well with those of Dr. Emerson, nor with the theory deduced from his table. They are given by Señor Don Manuel Trelles, chief of the statistical department of Buenos Ayres for the year 1868, and include the period of the terrible cholera plague which raged in that city during the first quarter of the year in question :—

Born of Native parents,	1,230 boys,	943 girls,
,, ,, Foreign ,,	1,480 ,,	1,404 ,.
,, ,, Mixed ,,	518 ,,	521 ,.
Total,	3,228 ,,	2,871 ,,

From these figures it appears that among the natives—the greatest sufferers of all—*four* boys were born to 3 girls, or a male excess of 30 *per cent.* as a total for the year; whereas Dr. Emerson only found a 17 *per cent.* surplus of girls for the cholera months, and that in the infected districts only. In one of the worst affected Buenos Ayres parishes —Piedad—*two* male births occurred to *one* female.

Yet in the face of such a manifest contradiction we are asked to believe that during or just after an epidemic, Nature produces females with all possible alacrity to avert the calamity of the species becoming extinct! Thousands have eulogised this "admirable provision." It should be borne in mind, however, that few women are capable of becoming mothers twenty-five times, whereas the power of procreation in the male, from a physiological point of view, is virtually without limit as to number of offspring. Hence when the race seems in danger of extinction, females, these theorists intimate, are in the estimation of Nature at a premium, and males at a discount!

But surely when statistics show that this surplus does not always occur under such circumstances, it should make us hesitate before asserting it as a law of nature. Moreover, I must take this pretended "natural law" to task for one thing,—it seems utterly to ignore the marriage rite as established by all Christian nations at the present day.

It is well known that throughout Christendom there is a considerable surplus of females in the adult population. Now if, when pestilence carries off 1 or 2 *per cent.* of any community, Nature, to insure the perpetuity of the race, somehow decrees an additional over-supply of 10 or 20 *per cent.* of females, what can she mean but polygamy?

The absurdity of such a statement as that nature really does what she is thus represented as doing, by those who advance such theories and problematical explanations, is surely apparent.

Nature is next credited with having so arranged matters that an excess of male children must die during infancy, or during the first few years of life. Darwin states, " During the first four or five years of life more male children die than female; for example, in England during the first year 126 boys die for every 100 girls, a proportion which in France is still more unfavourable." And then an explanation is offered of this excessive mortality. Dr. Crichton Browne imputes it to the larger average size of the male at birth, which renders it more liable to injury during parturition. Thus it is stoutly asserted that this excess of males at birth is not only perfectly natural, but is a special provision to meet this excessive mortality, which is also natural, and further, to provide for the contingencies of war, famine, and accidents, to which men are especially if not exclusively exposed. But on this hypothesis the same excess of males ought to obtain among the lower animals, for the males must be larger at birth, and they are exposed to danger far more than the females.

What is the fact, however? Darwin himself states that among racehorses the proportion of births is 99·7 males to 100 females, and amongst greyhounds 110 males to 100 females, but among dogs generally he considers this proportion exceptional. With Cheviot sheep the proportion is reversed—97·9 males to 100 females, and, taking England and Scotland together, 97·7 males to 100 females. This author's preconceived idea, however, makes him say that "probably this would not hold good at birth"—the figures representing the proportion a few months after birth. With the polygamous animals the females must be in excess, if, as the same author states, one male often has a herd of 100 females.

But in spite of this asserted persistent and universal excess of males at birth, it seems that Nature has made an erroneous calculation, for at the twentieth year there is a considerable preponderance of women. Thus, to quote Mr. D. A. Wells again, "To illustrate the changes in the numerical relations perfectly, take the following example from two United States Census Reports: In one decade, under 5 years of age, male excess $5\frac{1}{2}$ per cent. The next decade, under 5 years, male excess 4 per cent.; from 15 to 20 years, female excess 4 per cent.; from 30 to 40 years, male excess 14 per cent.; at 70, females predominate 5 per cent.

"The numerical law of the sexes then is this: 1st, There are more males than females born by about 4 per cent. 2nd, At twenty this preponderance is entirely lost, and there are more females. 3rd, At forty the balance is again the other way, and there are more males. 4th, At seventy the sexes are nearly even."

That is, from twenty to forty, the period when marriage is most generally contracted, there is a considerable excess of women who are thus doomed by nature to celibacy, until, presumably by reason of the additional dangers of maternity, the balance is restored, though only for a time. But if the preventible wars and famines are needful even to attain this balance, it would seem that as civilization progresses there will be an increasing preponderance of males, and that therefore celibacy, or polyandry, is the highest state to which the race is destined by nature. Surely again the wrong path must be traversed when inquiries lead to such conclusions as these. The facts are indisputable, and therefore it must be that an erroneous interpretation has been applied to them.

But again, another attempt is made to explain the excess of females, at least in Oriental countries. Thus Darwin, in the *Descent of Man*, says: "It has been supposed that with mankind polygamy leads to the birth of a greater proportion of female infants; but Dr. Campbell gave special attention to this subject in the harems of Siam, and he comes to the conclusion that the proportion of male to female births is the same as from monogamous unions." Darwin further states that English racehorses are highly polygamous, yet the sexes are born in about equal numbers among them. Individual mares, however, in some instances, produce nearly all of one or the other sex; he mentions an Arab mare that, with different sires, had seven fillies. Of the 25,560 births recorded among this stock in England during twenty-one years, the excess of females born was less than one in 600, or 99·7 males to 100 females. Here again, a contradiction is presented; but a point has been gained by Darwin's tacit assumption that *the law ruling sex must apply to other animals as well as to man.* Presumably, then, all the theories and assumptions hitherto discussed are false and contradictory, because they rest on too narrow a basis. Let but the view be extended a little, and it will be found that the "natural tendency to a preponderance of the male sex at birth" will give place to a more rational hypothesis; for on a closer examination of facts it will be clear that there is no uniformity whatever in the matter, save a semblance to rule in a few countries closely allied in race and habits.

The statistics already cited from Mr. Wells show that so far from a uniform male surplus of 4 per cent. in the births, as the writer claims, in some countries the excess is probably of females. The figures which he presents are only approximately correct, as will be seen by comparing them with the columns from which they were taken; and if several censuses are compared, there will be found such never-ending variations, that both the curious and interesting features of this "natural law" rapidly vanish. Furthermore, Mr. Wells did not take his figures from the "aggregate" column, but from a selected class—the "Native Whites." Had he taken the trouble to examine the selected classes of the adjoining parallel lines, he would have found very different results. The statistics of the Negroes, and more especially of the Mulattoes, show a large *excess of females* at birth; the latter giving a surplus at every selected age even up to one hundred years. Most races and nations, in fact, differ in this respect. The Jews, for example, as already stated, have an unusually large proportion of male births. In the several Prussian States, the excess ranges from 12 to 44 per cent. The redundancy of females among the negroes and mulattoes—which varies from 5 to 1 per cent.—is an interesting fact that I have never seen commented upon, but this, like most of the other difficulties presented, will be explained further on in the course of this inquiry.

The following extract from the *Globe* (London) of March 1882 is eminently suggestive :—

"The tendency of the two sexes to equalise their comparative numbers is well attested by the returns of population in the Australian colonies since they have begun to increase in material prosperity. Twenty years ago the preponderance of male over female colonists was exceedingly large, the totals being for the whole of Australia 737,000 men against 527,000 women. The difference now is rather less, the totals recorded in 1879 being 1,500,000 against 1,216,000. It is, however, somewhat remarkable that during the latter portion of the time under consideration this equalising process has not been going on with anything like the rapidity noticeable before.

In fact, the tendency was rather in the opposite direction; for whereas the proportion of men to women in 1861 was about 10 to 7, and in 1871 about 10½ to 8½, in 1879 it was nearly as much as 10 to 8. But in the case of the colonial districts which are improving most rapidly, the assimilation is still going on with full force. Thus in New Zealand, which shows a very high rate of increase in general population, the proportion of women rises from 38 to more than 41 per cent., whereas in the older and more settled districts, where there are larger towns, it remains almost stationary." This too will be explained later on.

Thus it is evident that there is no uniformity in the numerical predominance of either sex, and one step further may be taken, enabling me to say that Nature tends to restore the balance when it has been disturbed, and this points to a tendency on her part to equality, rather than towards male excess. For instance, M. Guillard, in his *Statistique Humaine*, shows that in the times of Louis XIV. and XV. there was a scarcity of men as compared with women; yet at the beginning of the Revolution the proportion was restored. Then, the wars of the Revolution and the Empire caused such a deficiency that in 1820 there was a lack of 868,325 men.

In 1835 this difference was diminished to 619,508.

In 1840 ,, ,, ,, ,, ,, 420,921.
In 1845 ,, ,, ,, ,, ,, 316,322.
In 1850 ,, ,, ,, ,, ,, 193,252.

From this it would appear that Nature's plan, in the case of a desolating war, is to strive directly to make up the deficit. But when pestilence comes,

there are those who affirm the absurdity that
Nature sets about producing females, irrespective
of the sex smitten—which is chiefly the male; thus
aiming to replenish her numbers in the course of
several generations.

There is also a difference between the percentages
of male and female births in city and country; statistics show a male excess in towns. There is also
a larger proportion of females in illegitimate than
in legitimate births. The careful compilations of
Belgium show that the price of breadstuffs affects
not only the fertility of the people, but also the
proportion of the sexes, in a slight degree.

Thus it seems that Nature has some plan of
reciprocal adjustment by which she maintains
substantially an equilibrium, though at times permitting local or temporary excess or deficiency.

A further inference, however, from the foregoing
statements is that Nature tends universally to
produce both sexes equally, and to restore the
equilibrium when it has been artificially disturbed;
and if the view is extended this will become abundantly evident, as may be gathered from the above
quoted statistics, which might be indefinitely
multiplied.

The only admissible explanation of the existing
state of things seems to be that some external
disturbing causes exist, varying with different
countries and circumstances, by which Nature's
original equality is from time to time unsettled,
but that, as seen above, wherever the disturbance
is great, Nature's power is exerted to make up

the deficiency, and to restore the equilibrium.
And that this must be so is evidenced by the
broad fact that, taking the whole world, the numerical proportion of the sexes is nearly equal; for
although there is a preponderance of adult women
among European nations, and a preponderance of
males at birth, this excess does not grow larger, but
rather the reverse, as return to a more natural state
of society is made. Moreover, the excess does not
increase even under present conditions; which is a
proof that Nature is averse to inequality, and works
against any preponderance of one sex.

But the differences in the relative proportions of
the sexes in different nations, in town and country,
in legitimate and illegitimate births, surely all tend
strongly to suggest the existence of variable secondary causes affecting the balance. The excess of
females in Europe is quite explicable on the assumption that the production is originally and naturally
equal, but that artificial and largely preventible
causes, such as wars, famines, accidents, immorality,
and social conditions, have disturbed the balance in
a large number of cases. And I shall be able hereafter to explain somewhat more fully the operation
of some of these causes in producing local inequalities.

If, then, the preponderance of either sex in society
at large, or in any particular nation, is due to
secondary causes, we may hope to discover these
causes; and, arriving at a knowledge of the laws
governing the production of sex in the *aggregate*,
might apply the same to the case of a *single family*.
But as a matter of fact the reverse mode of

reasoning is the scientific one; namely, of observing what happens in the family, and generalising therefrom to the nation and to the whole race; and this indeed is the rule which I have pursued in making my investigations, and which I shall follow in the present inquiry.

The neglect of this rule hitherto has been the bar to success. Some have started with theories, and have become disheartened on finding that they did not harmonise with the facts; and though a few have made observation the basis of their study, their range has been too confined, and their theories have been constructed on too narrow a foundation of facts, to render them of any real value. Still these writers have often arrived at partial truth, when once they have got rid of the bias resulting from the idea that Nature favours inequalities of production. Thus, Dr. Napheys, of Philadelphia, the author of *The Physical Life of Woman*, which has been reprinted in England, makes the following statements, which I fully endorse so far as they go:—

"The quantity and quality of food; the elevation of abode; the conditions of temperature; the parents' relative ages, mode of life, habits, rank, etc., have all been shown to be causes contributing to the disproportion of the sexes."

But it is requisite to have more than this if a hope be cherished for any solution of the mystery of sex. A foundation as broad as creation itself must be found upon which to rest an enduring theory. Any of the theories now current can be undermined at pleasure. The theory which so many accept, for example, that the parent actuated by the stronger love at the time of conception gives his or her sex to the child, is a very harmless assumption, and in fact possesses several advantages. It inculcates love, and it is equally impossible of proof or disproof,—for who shall attempt to gauge accurately another's affections at any given time?

But suppose love is assumed to be the controlling principle, the embarrassment which thereby presents itself is at once apparent. Newton and Butler both agree that *Analogy* is the supreme rule in science; and if love—whether abstract or concrete, Platonic or sensual—be the determining factor with one race, must it not be the rule with all? And why not with the animals as well? But what decides it when no love exists between the parties, which is presumably a most frequent occurrence?

The idea of animals possessing any genuine love, apart from an instinctive maternal yearning, is sometimes disputed; yet eminent naturalists say that even insects, as well as the whole feathery tribe, are seen to display their charms to the best advantage during courtship. It is given on high authority that the beautiful male colours of most species are given for the express purpose of winning the affections of their mates.

Let it be conceded, however, for the sake of argument, that sex, throughout the animal kingdom, is dependent upon an excess of affection in one of the procreating parties. But with this concession, it is by no means clear—to take a single instance—how the principle works in the case of those fish which deposit their spawn and leave it to the chance of being fertilized by some passing male.

Moreover, it has recently been discovered, by repeated experiments, that the sex of butterflies depends upon the surrounding conditions of the larva, or caterpillar, rather than upon the anatomical structure, as has heretofore been universally supposed. The sex of almost everything else in Nature is determined at the time of conception, or very early in the embryonic stage; and why should these creatures form an exception to the rule?

Caterpillars that were fed abundantly for a time before entering the chrysalis state, came out female butterflies in the proportion of over *ten* to *one*, while those which were deprived of food altogether, came out with a like predominance of the male sex. M. Gentry confirmed this fact in his experiments with moths.

Mr. Thomas Meehan, of Philadelphia, has shown that in plants, the highest scale of nutrition gives the female (flower) sex, and the lowest scale, males. Botanists claim that an excess of heat over light produces male plants, and the reverse, female. Furthermore, it is proved that a large part of the vegetable kingdom is hermaphroditic—containing both stamen and pistils in the same calyx. In the animal kingdom, also, this is the case. Some of the lower crustacea are able to propagate their kind asexually, without distinction of sex; hence, scarcely any genuine males are to be found among them. But with certain *cirripedes*, and perhaps with certain kinds of fishes as well, it seems to be established that more than one male is requisite to impregnate the female, and hence a great disproportion in the respective numbers of the sexes naturally follows.

The sea-worm, emerging from the egg a neuter, grows to many segments; then a perfect male or female forms itself out of one portion and drops off from the neuter parent to continue its life elsewhere. Other phenomena equally striking might be adduced, but enough has been presented to justify the conclusion, that no theory of the law of sex can be worthy of serious consideration, unless applicable to the animal kingdom in general as well as to mankind, it being unreasonable to suppose that each species is governed by a distinct sexual law.

But I have as yet only considered what the schools would call *objective* difficulties—pertaining to the matter discussed; and there remain those of a *subjective* nature—inherent in the investigator. It is perhaps impossible for the human mind to pursue without bias any train of thought; like the slender vine or creeping ivy, it seems to be a part of its nature ever to cling to, and depend upon, some other body or thing. As soon as one object is removed the tendrils are stretched forth to grasp another, and interests, preconceptions, prejudices, passions, affections, and education, all help to influence the judgment.

And here it may be well to call attention to the nature of the evidence which should be sought in investigating the law of sex.

Beyond all others, mathematical truths admit of a positive and exact demonstration. Metaphysical conclusions are reached by processes of systematic reasoning, while doctrines of the physical world must have *facts* for their basis, in order to inspire belief. Historical facts rest chiefly upon the testimony of witnesses, and are relied upon in proportion to the number, credibility, and unanimity of those witnesses. The evidence required to produce conviction in the question now under consideration is simply *facts*, and such chiefly as every person can verify for himself. Historical data, from the very nature of the case, must be received on the declaration of others, but the phenomena of sex are constantly observable everywhere, and whoever presents statements, the truth of which rests wholly upon another's veracity, should be listened to with caution.

In order to treat the subject methodically, the question here arises, where is it best to begin? Considering the somewhat exclusive design of this volume, probably all will readily agree that Nature's masterpiece, MAN, should be the first object of scrutiny, rather than primordial germs, the growth of plant or insect life, etc. At the same time, I believe that the theory advocated in this work is applicable to the lower organisms equally with man, and herein, I maintain, lies its strength. If it could not be thus applied, it would be worthless as a complete theory.

But mankind is the chief subject under study, and the great mass of evidence collected is made up of facts such as every reader can easily gather and examine for himself, without taking the voucher of any one for their correctness. At the same time, phenomena are brought in from many other sources as collateral testimony. Reliable statistics—compiled not in the interests of any league or theory, but by a government, for the general welfare of the nation—stand unique as tests for any theory.

Although the mass of evidence is immense, none need despair. With patience, and study, order will gradually appear amidst the chaotic accumulation of facts, as our enquiry proceeds; and if there are still found, as is sure to be the case, a number of stubborn facts that will not adapt themselves to any theory, it will be necessary to remember what the Rev. Dr. Newhall very pertinently says:—"Not until we know all things, can we see all truth to be harmonious with itself. In our study of Nature and of history, after being perfectly convinced of truths, we always find ourselves carrying along packages of exceptional facts which we cannot at the time adjust to others, but which we are sure that we can, as we grow wiser, drop into their proper places;" and before he closes this volume, the reader's confidence will grow stronger as the circle of facts explained grows wider, and the refractory facts one by one settle into their places, helping to support the theory they have so long seemed to oppose.

CHAPTER III.

FALLACIES OF CURRENT THEORIES.

IN this chapter I do not propose to enter fully into
a consideration of the very numerous theories
which have at one time or another been held in
estimation, but merely to discuss some of those
which have obtained a sufficient credence to entitle
them to notice; first, because this will aid us in
grasping the problem, and will bring out the points
which especially need explanation from the contradiction they give to every extant hypothesis; and
secondly, because, while clearing the ground of much

obstructive matter, it will place the reader in a position to estimate accurately and immediately the value of the theory I herein discuss.

There is no need to enter into any details concerning the extravagant fancies which have been entertained by ancient writers on this subject, and indeed I have not space to insert half their vagaries.

As might be expected from the number of investigators mentioned in the table compiled by Burdach (Chapter I.), extraordinary theories were accumulated, until, towards the close of the seventeenth century, more than *five hundred* were extant. Few of these, however, are worthy of serious consideration in this more enlightened age.

Owing, however, to the advances made during the eighteenth century in the sciences of anatomy and physiology, and in microscopic research, many new theories were started, several of which are conspicuous, viz., the Ovulary—making sex an inheren quality in each *ovum*, independent of outward influences, the father merely arousing its dorm nt powers; the Spermatic—making it wholly depen dent upon the seed of the father (Galen was one of the spermatists, believing that the ovule is a mere bed for the precious germ, while Pythagoras and Aristotle were ovulists); the Syngenesian, the Epigenesian, and those of Metamorphosis, Pre-formation, and Post-formation.

Professor Huxley, in sketching the history of evolution in Biology (*Science and Culture*), states that Epigenesis was defined by Harvey, who adopted it from Aristotle, to be the "successive differentiation of a relatively homogeneous rudiment into the parts and structures which are characteristic of the adult." Here of course the determination of the sex was the result of one or more of these differentiations at a period subsequent to conception, as the primitive germ or rudiment is the same in both sexes, and thus sex was presumably determined by causes operating during gestation; and this theory would therefore seem to support the hypothesis of a uterine if not an ovularian origin of sex.

Metamorphosis, on the other hand, was the doctrine that the entire chick *as a whole* really exists in the egg previous to incubation (and the same *mutatis mutandis* for the mammalia, including man), and that the process of incubation or gestation is simply one of expansion or unfolding of organs already existing in miniature. It is merely growth, and thus change of form, but involves no structural development or addition. This is less distinctly ovularian, but the theory is not explicit on the matter of sex, or indeed on the subject of the origin of the miniature but complete body. It is not clear whether the whole chick, and therefore of course its sex, is conceived as existing *ab initio* as a part of the parent and before impregnation; but the inference is that the sex, like the whole body, is existent in the ovum before impregnation, and the theory of development of the embryo by metamorphosis seems thus to support the ovularian origin of sex, that is, that some ova are male and some female,—in which case of course sex at will would be impossible.

Syngenesis as a hypothesis had little to do with sex; it simply asserted the belief that each part assimilated various molecules *of like nature with it-*

self, which were in existence everywhere, among others, in the world as " organic molecules," and which it selected from the rest and attached to itself, thus growing, by addition of like materials, much as a crystal grows in a solution of sugar or salt by the selection of the molecules of sugar or salt, and the rejection of those of the water or other solvent. Here of course the sex was determined before syngenesian development was begun, and upon the origin of sex nothing definite was asserted.

The older theories of sex were then chiefly ovularian or spermatic, with various theories of development superadded, as those of epigenesis, metamorphosis, and syngenesis. Modern science has shown that neither of these theories is tenable, for the facts of heredity offer a direct contradiction to all of them.

Other theories more vague have long held possession of the popular mind, possibly because of their vagueness.

Thus the theory that it is the mother's will— or her imagination—which may be induced to determine the sex, has been poetically expressed in the following injunction :—

> " A boy you wish ?   A beauteous boy behold,
> With lips a cherry red, and locks of gold ;
> Like him for whom Alexis sighed of old.
> If female fruit you rather covet, view
> A heavenly Venus such as Titian drew."

It may be left for disconsolate parents with large families, all of the opposite sex to that coveted, to testify as to the truth of these lines.

I must now turn to more sober considerations in which some effort is made to establish the theory of sex on scientific, that is anatomical or physiological bases.   One of the most carefully considered of these theories, and one therefore that claims some examination, is that advanced by Dr. Napheys, already mentioned.   His theory is Ovularian in its form, *i.e.*, he maintains that the sex is determined in the ovum independently of the male or fertilising element, and he says that "science can now speak confidently in the affirmative," on this point.   He cites in proof M. Thury, a Geneva professor, who claims to have discovered the means of procuring sex at will.   The matter was brought to his notice by the oft-repeated statement that queen bees lay female eggs first; also that hens do the same—or are said to do so.   From hens to mares is but a step, and he finds that when mares are fertilized late in their heat, they have male offspring—accounted for on the hypothesis that the ovum passes through two distinct stages of imperfect and complete development.   Hence, if females are desired, unite early in the period, and late, if males.

In support of this theory Dr. Napheys quotes a certificate from a Swiss stock-raiser, dated February 1867, which gives twenty-two successive cases where heifers were obtained.   His herd was of forty-two cows of all ages : he testifies to having succeeded twenty-nine times and failed in no instance. Another reported success of this plan is found in the *Medical and Surgical Reporter* of Philadelphia. A " respectable physician," also in the same journal, writes that numerous confirmations of Prof. Thury's theory have come under his professional observation, and proved fairly correct.

Dr. Ira Warren, in his *Family Physician*, cautiously says, there is some reason to suppose that the time of impregnation governs the sex.

Should the puzzled reader ask me to give him a plausible explanation of the fact that so many witnesses can thus testify to their own similar experiences, and yet be deceived as to the truth, I must confess that I give it up. It is the psychological riddle which has perplexed me from the outset far more than my subject of inquiry. The witnesses appear to be credible, and it is very difficult to believe that they are deliberate falsifiers, or that they are self-deluded as to the actual result of their experiments. But that they are wrong in their conclusions I promise to convince every candid reader.

The following extract from the *National Live Stock Journal* for August, 1874, is interesting, as a comment upon the above theory, with the additional element of the relative age of the parents ; and the conclusion arrived at by a journal devoted to practical breeding and stock-raising is important :—

" It has long been a study among breeders to control the sex of calves and colts, and many have been the theories advanced, and experiments made with reference to it. Coupling at certain stages of heat, and mating animals of certain ages, etc., have been resorted to, and about the time some one has made up his mind that he has discovered the true rule, events have transpired to show that he is mistaken.

"Breeders have great faith in their power to develop, by mere breeding, any quality which they deem valuable, and it is not strange that they should have fallen back upon the fundamental idea and sought to control the sex, as they do colour, milking qualities, speed, form, etc., by selecting sires and dams coming from families in which the desired sex predominates. But this theory, it seems, will not work with any more certainty than the other, and we suspect that there is a general average of both sexes produced; and that in the case of a mare or cow producing a greater number than the average of one sex, there is a tendency in the offspring to depart from it in the other direction—at least far enough to restore the equilibrium. If this is not so, the facts seem to show that the tendency to produce either one sex or the other is not a peculiarity which is transmissible, and that breeding for females does not promise a large measure of success. As to the reproductive power of females from dams far advanced in years, which, while it may or may not be sustained by the facts, is certainly worthy of investigation, we should be pleased to hear the experience of breeders upon this point.''

It will be noted that practical experience here suggests the conclusions I have arrived at in the previous chapter, viz., that the sexes are produced equally, and that Nature tends to restore the equilibrium when any disturbance occurs.

And if the curious reader will inquire for himself of different breeders, he will find that one will very likely think well of this theory; a second will admit that at times he is almost led to believe there is something in it; while a third will express his contempt for the propagators of all such theories. This, in substance, has been my experience in consulting with breeders everywhere.

Upon the point in question, the testimony of Darwin—who made such a speciality of sex—ought to have great weight. In the revised edition of his *Descent of Man* he says: "The period of impregnation has been thought to be the efficient cause in determining the sex; but recent observations discountenance this belief."

In searching for this law of sex from a purely scientific standpoint, something worthy should be looked for, something in harmony with the simple and majestic laws of Nature, the grandeur and perfection of which are precisely in proportion to their simplicity and regularity.

The Thury-Napheys theory is at best but a surface theory; it does not pretend to give an explanation; it states facts, but lays down no rational bases for their support. Looking at it from the elevated point of view just mentioned, one is compelled to ask, Is it consistent to suppose that Nature has locked up this important principle in the *ovulum* of the female, decreeing that to-day, to-morrow, and next day, it shall be capable of producing one sex; after which, as if by some juggler's sleight, it shall be transformed into a germ of the other sex? Would it not be positive confusion if Nature has, as Dr. Hough maintains, given but six days, at most, per month, in which females can be conceived—and if in the last two or three of these they must be insufferable viragoes; and to the males has allotted as many more days—on the first three of which shall be produced specimens as much too effeminate as the preceding ones were too masculine? Is it possible that Nature is thus constrained to inflict misery upon so large a portion of the human family? Facts certainly do not corroborate this assumption by showing nearly one-half of the human race wrongly sexed. And further, how would this theory apply to twins, or to multiparous animals, where there may be several of either sex in one litter?

Now to all these difficulties the theory in question offers no reply, nor in my opinion can it ever offer a satisfactory one. It may, however, apply in the exceptional cases of females of delicate constitutions (see p. 177).

The next theory which I propose to consider may be termed the "Alternate Theory." It will at any rate show how easy it is to be misled by coincidences. It is simply this: That Nature makes all human *ova* either male or female, and supplies masculine one month, and feminine the next, thus preserving substantially a sexual balance. People are often successful in applying a theory such as this, or at least they believe so. Unfortunately, as there are but two sexes, by the law of probabilities just one half of the cases examined must be found in favour of almost any theory, while coincidence may increase the proportion in the limited number of cases examined to the extent of another 10 or 15 per cent. Now all facts that make for any hypothesis to which an inclination is shown, will impress the theorist much more strongly than those which clash with it; and hence, remembering the favourable cases better than those which are adverse, his conviction increases that the theory must be true. Were one to frame a theory that those born on Monday, Wednesday, and

Friday, are girls, and on other days boys, thousands of instances could at once be brought to sustain it.

I must now pass over many other theories which I have also examined and found to be fallacious, and come to an important and still very prevalent theory, which is as significant as it is popular, and which possesses three special features more or less blended by different authorities. It is important because it has so many positive facts which seem to favour it, and numerous writers, ancient and modern, have given it their support, declaring it to be " the one generally accepted." Hippocrates is claimed as the father of this theory. I dwell upon it, as it undoubtedly contains a good deal of truth, though the right explanation has been hitherto missed ; and also because it will help to render my own theory more readily acceptable.

The three distinctive features of this theory may be designated: *Comparative Vigour ; Relative Age ;* and *Nutrition.* In other words, the parent who is physically the more vigorous at the time of conception gives his or her sex to the offspring. The older parent, *cæteris paribus*, is assumed to be the more mature and vigorous, and hence imparts the sex. But without nutrition there is neither vigour nor life, and it is therefore considered safe to assume that the better nurtured will be the more vigorous, and thus determine the sex.

The eminent naturalist, Baron Cuvier, says: " Curious indeed are the experiments of M. Charles Giron de Buzareingues concerning the procreation of the sexes, both in the animal and vegetable kingdoms. He is led to believe that the sex depends upon the comparative vigour of the parents. To obtain an excess of female offspring the father should be young and ill-fed, while the mother should be of mature years and highly-fed. The order should be reversed to produce males."

The three phases of the theory are here consolidated, and certain it is that with this combination—however unnatural the conditions—marked and positive results can be obtained. M. Giron, at a meeting of the Agricultural Society of Séverac, France, so long ago as July, 1826, proposed to so divide a flock of some three hundred sheep into two parts, that a greater number of males or females, at the choice of the owner, should be produced from each. The proposition was accepted and the experiment made.

The half that was highly fed and had young rams produced females in the proportion of *three* to *two* males; the other portion was rather poorly fed, had rams of mature years, and produced *three* males to *two* females, substantially as was predicted. I shall have occasion frequently to refer to M. Giron, whose experiments—if not his conclusions—display the exercise of more common sense than can be found in most writers on this subject.

We may further quote on this point the following from the *Medical Investigator* of Chicago: "Whether sex is determined by nutrition or not, is what scientists are solving. The experiments of Mr. Meehan, Mrs. Treat, and M. Gentry, on plants, butterflies, and moths, show that a high grade of nutrition gives females, and a low grade males."

Again Quillet, an old author, presents the nutritive theory, in a somewhat modified form, in the following quaint lines :—

> " Males are the strength and glory of a race,
> And female issue viewed by some as base ;
> An error obvious, for good sense allows
> That sex is best to whom the other bows.
> But let us, leaving this debate, our theme
> Pursue, and tell how wives with males may teem.
>
> Those most are apt for males in whom there meet
> Most of male vigour and the vital heat ;
> This the learned tell us, and experience shows
> That manly thoughts to manly love dispose.
> A bold, a gen'rous, and an easy mind
> Assist the sex to propagate its kind ;
> That future hymens may not strive for boys
> In vain, nor covet heirs with fruitless joys,
> Reason directs, that in the choice of food
> The parents carefully prepare their blood.
>
> Rich meals for you, ye bridegrooms, Nature bids,
> And thrilling draughts to cheer your wishing brides ;
> Sufficient for the nuptial joys the vine,
> And lusty boys are got by gen'rous wine.
>
> And you, ye wives, who with your husbands join
> To pray for sons to prop an ancient line,
> At meals with sparkling wine rejoice your souls,
> And freely pledge 'em in the modest bowls.
> Give to these precepts in thy heart a place,
> And masculine expect thy promised race."

Respecting this theory I may again cite Darwin, who said : " It has often been supposed that the relative ages of parents determine the sex of off-spring, and Professor Leuckart has advanced what he considers sufficient evidence with respect to man and certain domesticated animals, to show that this is one important factor in the result."

All this testimony is far from being groundless; and undoubtedly some glimpses of the great law sought for are thus obtained, but that any one or all of these conditions constitute that law is quite untenable. These conditions do not provide any principle of self-adjustment, or balance, the absence of which is, of itself, fatal to any theory whatsoever. They explain none of the anomalies brought to light by the statistician; and generalization upon them is impossible, further than to state the simple fact that a few given combinations are found to produce certain results, in a *majority of cases.* No explanation whatever can be offered of the host of exceptions; and these require to be taken into account, and a theory to be satisfactory must explain them.

Any one interested in the subject may call to mind families with an excess of either boys or girls, and find, far more frequently than otherwise, that they are of the sex of the fleshier parent, no less than of the older one. Superficial observers may be willing to rest upon these facts quite contented, but the further one looks the more numerous do the contradictions become. M. Giron, who left no stone unturned in his search for this law, settled down—for want of a better theory—upon that of *comparative vigour;* but it did not satisfy him. He discovered many facts which were in direct antagonism with it.

The following is such a perfect disproof of the physical vigour hypothesis that it would be a decided omission not to insert it; and best of all, it belongs to a class of facts which we can see verified all around us every day of our lives : M. Girou obtained statistics from physicians, of *eighteen* consumptive mothers, who gave birth to *eighty-seven* children. Upon the basis of *vigour*, most of these should certainly be boys; but no, *seventy-four* were *girls* and only *thirteen boys—one-seventh.* Any reader can notice that such is about the usual proportion under similar circumstances. M. Girou estimates that of strong, athletic fathers, and feeble, delicate mothers, *twenty daughters* are born to *one* son. As these facts are fatally opposed to the only hypothesis he could devise, these statements by him show the commendable impartiality of the man.

I think it will be clear from the foregoing that physical superiority, or vigour, is not the sole or even preponderant determining cause of sex; but an examination of its influence upon the actual casting of the sex, together with the several other related phenomena, will be given in another chapter.

A corollary from this theory of physical vigour gives, as mentioned above, the difference of age of the parents as the determining cause of sex, it being presumed that the elder parent is the more vigorous because more mature. This idea, however, was mainly due to the fact that in cases of any great difference in the ages of the parents, the children are observed to be chiefly of the sex of the elder parent. Thus Charles Roberts, F.R.C.S., in an article in the *Lancet* of December 11th, 1880, says that " we have statistics to prove that there should be a considerable difference in favour of the age of the husband if male progeny is desired." And all kinds of statistics have been compiled to show some point in favour of this theory of physical vigour; as, for example, taking the proportion of the sexes in first births only, and also in the first five years of marriage as compared with the second, and so on; but apart from the fact that the statistics of early and late marriages have been included, and are thus vitiated *ab initio*, they fail to carry conviction owing to the number of contradictions thus brought out. Of these some idea may be gained from the above citations from Girou and others, but I shall be able to reconcile them by the true theory.

Some striking statistics—which are frequently quoted in support of this relative age theory—are given by Hofacker and Sadler, the important point brought out for the supporters of the theory of physical vigour being, that in each case there is an *excess of children of the sex of the older and more mature parent.* These figures show the proportion of male live births to every 100 females :—

Hofacker.

Father younger than mother .	90·6
Father and mother of equal age	90·0
Father older by 1 to 6 years .	103·4
„ „ 6 to 9 „ . . .	124·7
„ „ 9 to 18 „ . .	143·7
„ „ 18 and more . .	200·0

Sadler.

Father younger than mother .	86·5
Father and mother of equal age	94·8
Father older by 1 to 6 years	103·7
„ „ 6 to 11 „	126·7
„ „ 11 to 16 „ .	147·7
„ „ 16 and more . . .	163·2

This theory might be credited in the cases of small differences of age, but what shall be said when belief is asked to the statement that a man of fifty is so much more physically vigorous than a man of thirty-six, that if both were to marry wives of thirty, the older man would have an excess of sons in the proportion of 200 per cent., while the latter case will only produce sons in the proportion of 103 per cent ? Is the man of fifty so much the superior of the man of thirty-six in physical energy? Experience must surely answer this in the negative.

But this theory, like many others, though not the whole truth, contains some glimpses of the truth, though in a shape not to be easily recognised. It is, in fact, as Darwin says in the passage just quoted, " one important factor in the result." There are, however, many more factors, and it will be seen later on that the application even of this one has been wholly misapprehended, and the results have therefore been contradictory and wholly unsatisfactory.

A passing reference is further necessary to a theory still current among large classes—hardly as a theory perhaps, but rather as a sort of self-evident and natural fact. I allude to the theory of *Genital superiority* which supposes that the more passionate parent stamps the sex of the child. It is not strange that such preponderance of passion should be supposed to influence if not to determine the result, for it is noticeable that persons with excessive passions often have children exclusively of their own sex. Here another glimpse of the workings of the true law is obtained, but nothing more :

otherwise Nature would be convicted of putting a premium upon vice and licentiousness,—an idea which condemns itself. Besides, such a theory can make no pretension to anything like an equalising principle. An ascendency of power, once gained, would in a few generations, on this principle, produce the dominant sex exclusively. Thus this theory, like all those which do not introduce some principle of self-adjustment, is overthrown by a simple appeal to the broad fact of the present balance between the sexes throughout the world.

I must also refer to another theory, promulgated originally by Dr. P. F. Sixt, a physician of Erfurt, who nearly a century ago wrote a work entitled, *Mysteries of Nature concerning the Generation of Man and the Voluntary Choice in the Sex of the Progeny.* For some unexplained reason the MS. was not printed till a few years since, when the late Dr. Trall, of New York, obtained it, and after much time and pains spent in having it tested, satisfied himself of its truth, and gave it publicity in book form.

Briefly it asserts that the organs of the right side, in both sexes, are masculine, so to speak, and those of the left side, feminine; that the secretion of one side cannot fertilise the ovum from the opposite side, and that only when the opposite elements from the same side meet can there be any offspring. Evidence was collected, experiments made with animals, and even Aristotle and Hippocrates quoted in support of the theory. All difficulties were explained away, and most circumstantial accounts of cases were given. Yet for a century the idea had remained dormant, and after its promulgation by Dr. Trall, and in a modified form by M. de Ferrendi in the *Scientific American*, Dr. Trall was at last constrained to publish some private letters from friends of scrupulous veracity, who, after repeated tests with animals, pronounced the affair " an audacious humbug."

As an example of an extreme phase of the Ovularian theory, I may in this place quote from the work of M. A. Debay, *La Venus Féconde*, which is also curious as asserting power in the parent—the mother only, in this case—to determine the sex of her offspring by a species of physiological discipline. From an analysis of the ovum he concludes that the whole question rests with it, the male element merely fertilising. He considers each ovum to contain a male and female principle, and that one or the other of these predominates according to the temperament of the mother. If she be very effeminate, the ovum will have the female element in excess, and will result in a girl. If the mother be something of a virago or of a masculine temperament, the male element will prevail and produce a corresponding result. He therefore suggests that a mother may by transforming her own organization affect the sex of her offspring, and he gives specific directions how to effect these temperamental transformations. This theory, inasmuch as it completely ignores the influence of the father, belongs to the ovularian theory of the last century, and is quite contrary to modern experience.

One other theory deserves a passing consideration, not for its own sake perhaps, but for the hint of truth contained in it. It is not formally enunciated as it stands, but seems to have originated with Professor O. S. Fowler, who, in some of his works of thirty years ago, declared that such was the natural law. His idea was that electricity effected this result—man being always positive and woman negative; but I believe he has never brought proofs to sustain his views. In a later work—*Sexual Science*—he states his convictions, after half a century's study upon this point, without, however, advancing any theory or facts, and says that the parent which casts the gender, originates the *opposite* sex. That is, when the father is the determining parent the child will be a girl, and most like him; but when the mother, it will be a boy, and most like her. This theory is closely analogous to another theory long in vogue, that strongly masculinized males have most daughters, while strongly femininized mothers conceive more boys, strongly resembling their mothers; and also with the vague notion current in the popular mind, that the father produces girls, and the mother boys. This idea probably originated in family resemblances, or possibly in a love of the paradoxical. In any case, these theories, like most others, contain a germ of truth, as will be abundantly shown in the sequel, when they are viewed in the light of the theory which I shall set forth in this volume.

The following as an interesting example of the futility of theories not based on a wide foundation of observed facts may not be out of place. An American physiologist claimed to have made the discovery that the heart of the female foetus beats with much greater rapidity than that of the male,

and that, consequently, interested parties might know
the sex of a child months before its birth. Many
experiments seemed for a time to point to the reality
of the discovery. More than fifty authentic and
successful predictions had been made, based upon the
notable difference between male and female fœtal
pulsations. At length there seemed to be a little less
uniformity in the published results than at first. A
considerable margin had to be left for failures in the
predictions ; some even expressed their disbelief in
the whole theory. But the final collapse came with
the testimony of Dr. Wm. E. Ford, of Charity Hos-
pital, New York. In endeavouring to ascertain the
sex of children before birth, he found all approxima-
tions even futile, save those based upon the comparative ages of the parents, which in numerous
instances afforded some criterion when the disparity
in years was great.

I have now probably mentioned most of the sexual
hypotheses which still obtain any credence. It will
be seen that none fulfil the requirements of the
case. They do not explain the difficulties presented
by the statistician ; they give us no clue to the
principle of self-adjustment which inseparably clings
to the law of sex ; few, if any of them, seem worthy
of nature ; many of them are palpably trivial and
absurd ; and it may be well, before closing this
chapter, briefly to gather up a few of the threads
running through it, so that a more definite idea may
be gained of the difficulties to be met, and the
requirements to be satisfied by any theory claiming
to be really authoritative.

First, then, the Ovularian and Spermatic theories fail to account for the production of twins of different sexes, and for the undoubted fact of inheritance of qualities and features from both parents in varying degrees, as well as of qualities from the parent of the opposite sex.

In the next place, the theories which refer sex to the time of impregnation fail to explain the existence of families almost exclusively of one sex, and the fact of both sexes being constantly found in the same litter of animals.

The Physical Vigour and Relative Age theory seems to explain a great deal, but contradicts itself when applied to experimental results.

Finally, the Genital Vigour theory is contradicted by the approximate balance of the sexes in the world, and fails to account for sex in the case of the lower animals, such as fish, mollusca, etc., where the question of passion cannot apply.

Any new theory to be accepted must therefore meet all these difficulties, and at the same time account for the preponderance of male births in Europe, of females among mulattoes and other hybrid races, as also among polygamous animals; and for the equality of the sexes among other animals. More especially it must suggest some principle of self-adjustment, by which not only is the balance of the sexes nearly preserved on the whole, but by which also, in cases of special disturbance, the balance tends naturally to readjust itself.

CHAPTER IV.

THE NATURE OF SEX AND SEXUAL EQUALITY.

Male and female the exception rather than the rule in nature—Both plants and animals afford evidence of this—As also do fishes—Each sex the equivalent of the other—Special organs met with in the higher forms of life—Division of function is the origin of sex—The propagation of trees and insects—Heterogenesis and parthenogenesis amongst moths, polyps, and bees—Alternate generation—Traditions of a bisexual progenitor in heathen mythology—The human embryo—The relative importance of the sexes—Man generally considered physically superior—Woman only an undeveloped man—The development of women arrested by procreative functions—Female plants less highly developed than male—Development antagonistic to reproduction—Man's susceptibility to disturbing causes—Contradictory opinions on the original sex of the embryo—Arguments in favour of female superiority—Females live longer than males, and are less liable to disease—The female amongst animals generally superior—Law of equilibrium as regards sex—The sexes throughout creation true equivalents—Plato of this opinion—Can the sexes be equalized numerically by natural selection?—Darwin's hesitation on the point—Fertility may be hereditary, but not the production of a certain sex—No equilibrium possible in such a case—Genealogical records contradict the theory.

BEFORE enlarging upon the question of sex at will, it is indispensable to consider the significance and, if possible, the origin of the sexes, and their comparative equality.

The existence of separate individuals in whom the two elements necessary to reproduction are fixed, and whose union is therefore necessary before fertilization can take place, is rather the exception than the rule throughout nature.

Professor Huxley, in treating of Biology, says: " Sustentative, generative, and correlative functions

in the lower forms of life are exerted indifferently,
or nearly so, by all parts of the protoplasmic body.
. . . Generation by fission and gemmation are not
confined to the simplest forms of life. Both modes
are common, not only among plants, but among
animals of considerable complexity. Throughout
almost the whole series of living beings we find
agamogenesis, or non-sexual generation. . . . The
phenomena which living things present have no
parallel in the mineral world."

In the vegetable kingdom, the male and female
organs are situated usually in the same flower. On
rarer occasions it is found that some flowers are
strictly male and others female, but even then both
flowers commonly occur on the same plant, and it is
only very rarely that exclusively male and female
plants are met with.

Among animals, all the lowest forms, and many of
the higher, including sometimes even fishes, are propagated asexually. Here, therefore, as in the vegetable kingdom, there can be no question as to the
superiority of one or the other sex; each part of
the organism shares in the strength or weakness of
the whole.

But as the development of life is traced to higher
forms, the functions are gradually specialised, *i.e.*,
special organs are set apart for the performance of
special functions; such as those of prehension, digestion, respiration, locomotion, and, in some cases, reproduction. Instead of generation by simple fission,
or germination from any part of the body indifferently,
special organs of reproduction are appointed and fitted
for exercising this function, and for nothing else.

The division into male and female organs is a later and higher stage; and the separation of these, and the lodging of them in different individuals, is a later and higher stage still. *Division of function*, then, is the origin of sex.

In a recent work entitled "*Sexes throughout Nature*," a view is put forth which may not be altogether unreasonable: "The first rise of a division of function must be everywhere accompanied by a corresponding evolution of organs ; yet the formation of differentiated structure is always expensive and to be avoided except for some sufficient reason. . . . Sexless offspring, sent out at less cost to the parent, can thrive and multiply with equal advantage to themselves, so long as there is abundant food and warmth ; but with any disturbance of the equilibrium, provident Nature, alarmed for the result, begins at once to circumvent the evil by a division of the reproductive process ; the result is a brood with more perfect endowments, which can continue the reproduction of the race, either asexually or sexually.

" The methods are various when the ends to be attained are the same ; but apparently among all those beings that have the two modes of propagation, any disturbance of equilibrium, from whatever cause, produces an immediate resort to the more expensive, but at the same time the higher and more effective process of sexual genesis. Thus a tree may be made to produce seed by under-feeding or by over-feeding, by various modes of disturbing the roots, or by pruning the branches, or merely the leaves, or by slightly girdling the bark. Unsettle the even balance in any way, and, in instant alarm, all the resources of the community are turned to the propagation of a young colony which can be sent off to continue the race elsewhere.

"The instinct of the bee teaches the same lesson. The semi-sexless members of the community can be developed at less cost of nutriment, and their division of work can be performed to advantage if there is a development of no conflicting interests; but with the death of the queen, the community is unbalanced, and is in danger of destruction. Some of the young that would have remained rudimentary in sex to the end but for this derangement of function in the body politic, are immediately matured into complete young queens—a process readily accomplished by giving them a generous and more stimulating diet, and a royal cell in which there is room enough to allow of the necessary growth and development. No one can fail to see the analogy between this process and the development of a leaf-bud into a flower. The annual autumnal destruction of the now useless drones is but a curious modification of similar vegetable economy seen in the annual fall of the leaf."

In the animal creation we have examples of *heterogenesis* (variable generation), and *parthenogenesis* (virgin generation). In insect life the moth generates larvæ, and these in turn become moths. The polyps, or coral-producing family, form communities and produce their offspring from themselves like the buds and ramifications of a tree; each species has its own way of branching, as much as different trees have theirs. With bees, the queens and workers are comparatively independent; the queen producing eggs in almost countless numbers from a single impregnation.

Alternate generation is well shown in the medusæ and other animals of low organization, where the sexually produced offspring gives rise after a time to a new and quite different form, asexually—by a variety of parthenogenesis. In some cases there are several steps of asexual generation, each producing a form differing widely from the preceding, until—the final product resembling the ancestor—sexual reproduction once more takes place, and the series is repeated.

The occasional appearance of a partial or apparent hermaphrodite in the human species might seem to give some shadow of colouring to the idea that man was once bisexual; and, indeed, the tradition of a bisexual progenitor—sometimes deified—is very widely spread. The Hindoo mythology, and that of Madagascar, may be cited, while the tradition is traceable even in classic times in Greece. And, curiously enough, it is now believed to be ascertained that at a very early embryonic period both sexes possess true male and female glands; hence, say Darwin and others, some extremely remote progenitor of the whole vertebrate kingdom must have been androgynous or hermaphroditic.

At this point it is necessary to consider the relative importance of the sexes, and to ask which is the superior. The opinions of the thoughtful in different ages should not be overlooked, but were the mere weight of authorities allowed to settle this question, it would probably be decided summarily by a large majority in favour of *male superiority*.

Those who maintain that the sexes are essentially equal are not numerous, while those who advocate feminine ascendency are indeed few. I will first notice some of the champions of man's pre-eminence.

It will naturally be understood that physical superiority only is here in question. From the earliest ages, it may be said that philosophers have contended that woman is but an undeveloped man.

Darwin's theory of *Sexual Selection* presupposes that a superiority has been involved in the male line and entailed chiefly on that sex; that men have developed muscle and brains much superior to those of females, and have transmitted their superior qualities chiefly to their own gender.

In treating of *Growth and Reproduction*, Spencer, in his usual characteristic language, says: "Genesis under every form is a process of negative and positive disintegration, and is thus essentially opposed to that process of integration which is one element of individual evolution." Women, he argues, are inferior to men because their development must be earlier arrested by procreative functions. As the demands of reproduction fall so much more heavily upon them, their higher advancement is correspondingly curtailed; hence they can never equal man physically nor mentally. Darwin scientifically adds to the male,—male evolution; and Spencer, as scientifically, subtracts from the female by his "earlier arrest of feminine development." Each of them, by different routes and new pathways, arrived at the same conclusion. Darwin, having with his indomitable patient industry accumulated a vast amount of evidence to sustain a certain hypothesis, felt entitled to speak authoritatively upon this point, in these words: "Thus man has ultimately become superior to woman," and feelingly added: "It is indeed fortunate that the law of equal transmission of characters to both sexes has commonly prevailed throughout the whole class of mammals; otherwise it is probable that man would have become as superior in mental endowments to woman as the peacock is in ornamental plumage to the peahen."

Dr. Hough, who is a thorough evolutionist, also favours male superiority, and presents numerous arguments to sustain his view. His position is: "That female plants, like female animals, are less highly developed than males, and are the result of an inferior developmental reproductive effort on the part of the female plants." He says all agree that "vigour" or development is directly antagonistic to reproduction; hence most females are begotten when the system is occupied in the process of growth, reparation, or disease.

M. Girou and Mr. Meehan come to a similar conclusion regarding plants, and Dr. Hough considers also that as man is a higher order of being, physically, than woman, he is more susceptible to all disturbing causes—which would explain the excessive male mortality already noticed. There are more males born idiots, deaf, dumb, blind, etc., than females, as seen by statistics; more idiots among men and more insane among women; more malformations by *defect* among males, and by *excess* among females; all which he thinks strengthen his case.

Girou also seems to consider man the superior, and thinks he finds young couples producing an excess of girls, and those of maturer years more boys.

Tiedman and others have advanced the hypothesis that every embryo is originally female, and that when arrested at an inferior point of organization it remains female—which harmonizes with much of the foregoing. Other physiologists, however, believe that the ova are originally neuter, and that the sex is determined by accidental circumstances at or about the time of conception. Others again, and among them Velpau, maintain that the original sex is masculine, and it is to be presumed they look upon the feminine as an offshoot from it, or as a modified and degenerate form of the true type of humanity.

Those who maintain *female superiority* have not such a formidable array of authorities to sustain them; yet, perhaps, the facts they adduce are as weighty as those presented by writers holding the contrary opinion. The general tenor of their arguments is, that woman is a later creation, and therefore of a higher order; that indeed she possesses functions—reproductive—in addition to anything man has; that her place in life is in every way superior to his. The following facts, they claim, go to strengthen this belief:—

Statistical tables show that females live longer than males: that they are less liable to diseases of almost every name—whooping-cough and diphtheria being among the few exceptions: seven men die suddenly to one woman—apart from violent deaths, which are in the proportion of four to one: infants die in the proportion of three males to two females, which proportion holds true even in still births: the deaths of males under one year are treble the excess of male births: animals generally show superior qualities in the female,—mares, for instance, are tougher than horses as a general rule, founded upon extensive observation.

As the majority of close thinkers and scientific investigators are, and always have been, of the sterner sex, it is very natural that they should be biassed in the study of this problem, and should solve it in the way most flattering to their own vanity. It is equally natural that few but women should advocate feminine superiority. But that the conclusions of both parties are altogether erroneous I am convinced, and I think that every attentive reader will be similarly persuaded before closing this volume. *Neither sex is physically the superior, but both are essentially equal in a physiological sense.* This is the chief corner-stone of my edifice ; the central, fundamental doctrine of this work.

That there exists a law of equilibrium in regard to sex, as in all Nature's operations, is a fact upon which I have insisted in the earlier chapters, and upon the strength of which the above uncompromising declaration is made. It is no mere *assumption* that there is a self-adjusting principle connected with sex, for it is a *fact*, as visible as it is indispensable and immutable. Its workings are to be seen on every hand as easily as the vivifying effects of the sun's rays upon the face of Nature. *It must be so*, or the sexes would not show the invariable tendency to equilibrium throughout the world which they now do, as already mentioned.

Let the reader understand, then, that the sexes in each species, from the lowest orders of animal creation to the highest, are herein claimed to be *true equivalents*; " equals, but not identicals, in development and in relative amounts of all normal force." That this is and must always be the case I think will become more and more patent with the perusal of each succeeding chapter. And there is sufficient evidence to sustain this view, even in the writings of those who advocate the superiority of one or other of the sexes, as the following instances will demonstrate.

Plato maintained that they were equal, although woman's social position in his day was far from encouraging any such conclusion. John Stuart Mill denied that there was any radical inequality between them, and even went so far as to claim for woman an absolute equality with man in all social relations.

Darwin himself says : " The sexes do not differ much in constitution before the power of reproduction is gained."

Dr. E. H. Clarke, who has become famous by his work on *Sex in Education*, says that the sexes are substantially the same at birth, diverge in middle life, but are again alike in old age. He says : " Neither is there any such thing as inferiority or superiority in this matter. Man is not superior to woman, nor woman to man. The relation of the sexes is one of equality, not of better and worse or of higher and lower. By this it is not intended to say that the sexes are the same. They are different, widely different, from each other, and so different that each can do in certain directions what the other cannot." He considers the nutritive and nervous functional capabilities of the sexes equal, not to say identical, but not so the reproductive.

Dr. Napheys says: " There is in fact a less difference between the sexes than is generally believed. They are but slight variations from one original plan. Anatomists maintain with plausible arguments that there is no part or organ in the one sex but has an analogous part or organ in the other, similar in structure, similar in position. Just as the right side resembles the left, so does man resemble woman."

The testimony of Professor Agassiz is quite to the point in this connection, when he says: " Generation is based upon the harmonious antagonism between the sexes, that contrast between the male and female elements that at once divides and unites the whole animal kingdom.''

Mrs. Blackwell, who has written a work mainly to establish this very point of equality, has presented the real issue in a clear and scientific form. She says: " The average males and females in every species always have been approximately equals, both physically and mentally. The extra size, the greater beauty of colour and wealth of appendages, and the greater physical strength and activity in males, have been in each species mathematically offset in the females by corresponding advantages—such as more highly differentiated structural development; greater rapidity of organic processes; larger relative endurance, dependent upon a more facile adjustment of functions among themselves, thus insuring a more prompt recuperation after every severe tax on the energies. The stronger passional force in the male finds its equivalent in the deeper parental and conjugal affection of the female : and in man the more aggressive and constructive intellect of the male is balanced by a higher intellectual insight, combined with a greater faculty in coping with details and reducing them to harmonious adjustment, in the female." The following is her way of stating the equation :

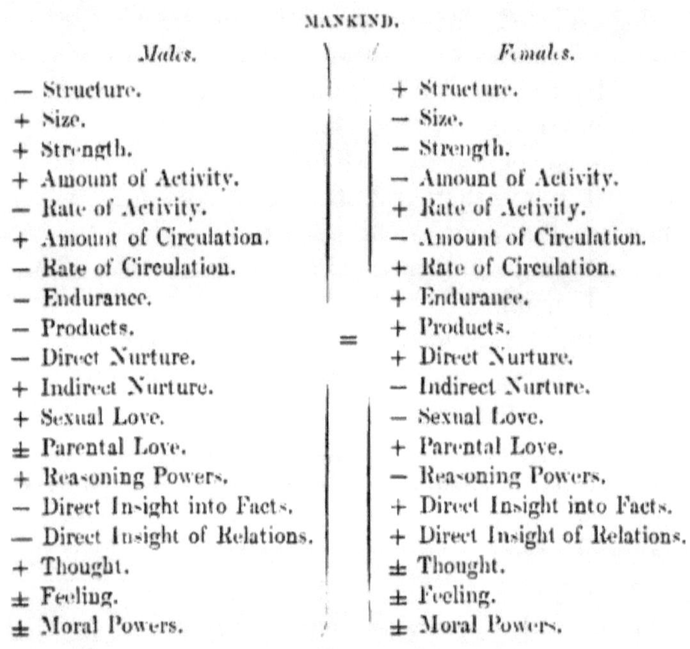

MANKIND.

Males.		Females.
— Structure.		+ Structure.
+ Size.		— Size.
+ Strength.		— Strength.
+ Amount of Activity.		— Amount of Activity.
— Rate of Activity.		+ Rate of Activity.
+ Amount of Circulation.		— Amount of Circulation.
— Rate of Circulation.		+ Rate of Circulation.
— Endurance.		+ Endurance.
— Products.	=	+ Products.
— Direct Nurture.		+ Direct Nurture.
+ Indirect Nurture.		— Indirect Nurture.
+ Sexual Love.		— Sexual Love.
± Parental Love.		+ Parental Love.
+ Reasoning Powers.		— Reasoning Powers.
— Direct Insight into Facts.		+ Direct Insight into Facts.
— Direct Insight of Relations.		+ Direct Insight of Relations.
+ Thought.		± Thought.
± Feeling.		± Feeling.
± Moral Powers.		± Moral Powers.

Result in this as in every other species—

$$\text{Sex} = \text{Sex};$$

Or,

" Organic Equilibrium in Physiological and Psychological Equivalence of the Sexes."

One more quotation from Mrs. Blackwell will suffice to make her position thoroughly understood; and I think few will even attempt to undermine it:—

"In women, if there is a greater arrest of individual growth than in men, the difference begins in the fœtal life; their comparative weight and size at birth are the same as at maturity; and if the former finish their growth earlier, it must be because relatively they grow more rapidly. The feminine circulation and respiration are both quicker; and so are the female mental processes. When the whole subject has been quantitatively investigated with sufficient exactness, I believe it will be found that what man has gained in 'massiveness,' woman has gained in rapidity of action; and that all their powers of body and mind, *mathematically* computed, are, and will continue to be, real and true equivalents. The premisses are already sufficiently known to compel me to this conclusion."

The one fundamental position taken up in this chapter—the equality of the sexes—is based largely upon the conviction that in no other way can a law of numerical balance be established. Although I have not yet arrived at the proper place fully to demonstrate how the equality of the sexes controls the numbers born of either sex, it is time to show that no tenable hypothesis has hitherto been devised to this end. Mr. Darwin, in fact, is the only one I have found seriously attempting it. Under the heading, *The Power of Natural Selection to Regulate the Proportional Numbers of the Sexes*, he argues as follows:—

"Let us now take the case of a species producing

from the unknown causes just alluded to, an excess of one sex—we will say of males, these being superfluous and useless, or nearly so. Could the sexes be equalized by Natural Selection? We may feel sure, from all characters being variable, that certain pairs would produce a somewhat less excess of males over females than other pairs. The former, supposing the actual number of offspring to remain constant, would necessarily produce more females, and would therefore be more productive. On the doctrine of chances, a greater number of the offspring of the more productive pairs would survive, and these would inherit a tendency to procreate fewer males and more females. Thus a tendency toward equalization would be brought about."

He maintained that the same course of reasoning was applicable to a redundancy of females, even in polygamous species. He also shows another way in which he considered that Natural Selection is able to balance the sexes. In case of an excess of males once more, he supposes—as there most assuredly would be—pairs with a suitable number of females, but a deficiency of males. This progeny would be stronger, and inherit from the mother a tendency to produce few males, which would conspire to preserve the numerical equality of the sexes.

Mr. Darwin himself, it would almost appear, had some doubts of the power of Natural Selection to equalize, by his announcing the theory interrogatively; perhaps he half feared that his readers would be unwilling to receive it. He grows bolder, however, as he proceeds, and thus speaks confidently at the close: "We may conclude that Natural

Selection will always tend, though sometimes ineffi-
ciently, to equalize the relative numbers of the two
sexes."

Now this plan, though plausible if viewed super-
ficially, seems to me a most insufficient one. As
to pairs producing fewer males than females, there
would be many such. I have accumulated the
statistics of hundreds of pairs of the human species
to test this point, as any one with sufficient interest
can likewise do; the usual excess of male births
among mankind exactly favours our object.

Families may be conveniently classified under
three nearly equal divisions: those with an excess
of male births; those with a surplus of females;
and those in which the proportion of the sexes is
about equal. The figures may be stated thus: 36
per cent. with a preponderance of males; 32 per
cent. with an excess of females; and 32 per cent.
of families equally divided. As there is so uniformly
a preponderance of several per cent. of male births,
this result is just what should be anticipated, and
it fortunately accords perfectly with Darwin's
Variability of Characters. Now for the application
of this plan.

Some pairs will produce a paucity of males,—32
per cent.; and as these produce more females, and
transmit a female tendency, thus, etc. But does he
forget that at the same time, as a counterpoise to
this 32 per cent. with its female proclivity, there
is a *thirty-six* per cent. with male excess, inheriting
and transmitting—according to his own theory—its
proneness to male superabundance? Thus, instead
of the male plethora being reduced, it must ever

increase at an alarming rate. In case of an over-supply of females, this method would be still more impotent to regulate the proportions than with male excess, especially among the polygamous.

Thus, the theory somewhat hesitatingly advanced by Darwin is quite insufficient, even when all his assumptions are admitted. He says that no one who accepts the evolution hypothesis can doubt that the offspring of families mostly male—or female either—inherit from their parents a tendency to produce that same sex, chiefly. But nothing is easier to prove than that there exists no such tendency. Fertility and many other qualities may be hereditary, but the production of a *certain sex* is not, nor could any equilibrium be preserved were such the case. I can cite scores of contradictions to this supposed tendency, without a single instance appearing in its favour; any reader who will take the pains to notice families around him will see that I am right. Historical proof can be found in all genealogical records, but the following evidence, respecting live stock, from Lord Althorp, must suffice for this chapter. That what his lordship says is correct, can be verified at any time ; he writes :—

" Are certain families more addicted to offspring of one sex than the other? Certain individuals assuredly are so, but is the tendency hereditary ? Clearly, not invariably on the sire's side. *Marmaduke*, for example, was a remarkable *heifer-getter* in his first season or two. *Moss Rose*, by *Marmaduke*, bred *all bulls* but one, and the exception was barren. Mr. J. G. Grove bred *Bustle*, of the Bliss tribe, and her produce was mostly (if not all) heifers ; yet

Bellona (her daughter) bred nothing but bulls, and *Princess Maude* (another daughter) all bulls, with one exception. Mr. R. Booth's *Windsor* begat Mr. Booth's *Lady Blithe* and Mr. Carr's *Windsor's Queen*. The latter bred all bulls; the former, eight or nine heifers and one bull. All the last-named four were of one tribe, and yet two of the dams recorded bulls only, and one, all bulls, with an exceptional heifer. And, besides *Lady Blithe*, *Stella*, (Lady Pigot's), a descendant of *Princess Maude*, had four heifers out of five calves. So that it cannot be said there is any invariable rule, or even assurance, one way or another. *Oxford*, 11*th*, and *Oxford*, 15*th*, were own sisters; the former bred bulls, whilst the latter gave rise to the numerous family of Holker; Sir C. Knightly's *Erminstade* bred all bulls at Barbraham till her last calf, but we did not find the male offspring of her sons more than usually numerous.

"Among all kinds of live stock kept in confinement and under artificial treatment, we believe likewise we have found male offspring to preponderate. The only fact we have ever observed like a law is, that if there be any unusual difference in age between the parents, especially when the female is by far the elder, the produce are for the greater part female. Individuals are found everywhere which have all male offspring; occasionally, but much more rarely, individuals which have all female offspring; but these offspring do not display the peculiarity of the parents to any marked extent. It would be interesting to have any authentic record on this head put forward. In no case would information be so valuable as from breeders of thoroughbred horses and pedigree short-horns, because, from the length of time during which these varieties of domesticated animals have been carefully registered, the instances which may be detected can be examined thoroughly, and the circumstances unravelled to an unusual extent.

"Our own opinion is, that though sires and dams are often to be found whose progeny are more of one sex than the other, no certainty exists that this progeny will differ in any respect from animals otherwise bred. But we have frequently found that disparity of age produces five-sixths female offspring. A five-year-old game hen produced to a cockerel of the previous season *eleven chickens*, and *nine* of them were pullets. An unusually old cow (as instances see Mr. R. Booth's *Modesty*, Mr. H. Webb's *Mayflower*, Mr. Foster's *Polly Gwynne*) generally finishes off with a heifer-calf. And we have noted that these last fruits of an old tree are generally abnormal in their tendencies; and not unfrequently, like twin-heifers, if they breed at all, breed only one or two calves, and these often turn out to be barren."

This evidence, in our opinion, disproves Mr. Darwin's theory of the hereditary tendency to produce a given sex. It is clear, then, that Natural Selection alone cannot readjust the balance of the sexes if there is a natural tendency in one of them to predominate. And yet it has been shown that on the whole there is no tendency in one of the sexes to predominate; for though there are districts and classes in which each in turn is produced in excess, yet these seem to neutralize each other, and if there is not an absolute equality in their numbers at the

present time, there is certainly no increasing, cumulative excess, as there assuredly would be if one sex
tended to produce in excess.

The conclusion lately arrived at on the consideration of the sexes physiologically is thus confirmed—
namely, that *each sex is the equivalent of the other*,
and that, taken generally, they are produced in equal
numbers.

Everywhere around me may be seen families, and
even whole classes, in which one sex predominates
very largely—sometimes to the exclusion of the
other. It is necessary therefore to admit that in
these cases at least the sexes, in the parents, are
not equivalents. Indeed, in no given case can they
be equivalents *absolutely*, or a hermaphrodite would
result.

To ensure sex at will, it therefore remains to
consider what inequalities really exist, and how
they are produced. In the ensuing chapters, therefore, I propose to enter fully upon this investigation.

CHAPTER V.

HEREDITY AND STERILITY.

COMMON observation goes to prove that children generally, more or less, resemble their parents, and science has given endless proofs of the inheritance of special qualities, and even of chance peculiarities reappearing generations after the original progenitor has passed away. And yet there is some uncertainty on the subject, for Mr. Francis Galton, F.R.S., has found reason to write a work, with a mass of most carefully compiled statistics, to show that genius, on the whole, tends to be hereditary. And Darwin makes the following statement in the *Origin of Species* :—" The laws governing inheritance are for the most part unknown. No one can say why the same peculiarity in different individuals of the same species, or in different species, is sometimes inherited and sometimes not so; why the child often reverts in certain characteristics to its grandfather or grandmother or more remote ancestor; why a peculiarity is often transmitted from one sex to both sexes, or to one sex alone, more commonly, but not exclusively to the like sex." It seems, therefore, necessary to devote a few pages to examine how far the features, disposition, and peculiarities of a parent may be looked for in the children—more especially as this will help us in determining what are the causes which influence sex.

Great attention has been recently given to education; it is looked upon as a sovereign remedy for crime and many other diseases of the body politic. But probably the most urgent question of the times is this: Is not *generation* of more consequence than *education?* None can fail to see the incomparable advantage which the preceptor of the child Moses would enjoy compared with one who should attempt to guide aright the stripling Ishmael. Were Socrates and Nero alike in infancy? or St. Vincent and Pope Alexander VI.? Lady Jane Grey and Catharine of Russia? Was it education merely that made the contrast between Newton and Boswell? or Henry Kirke White and Edgar Allen Poe? between Howard the philanthropist and Robespierre the human fiend? In improving the breed of domestic animals, is the best attention given to the training or to the blood?

It has been truthfully remarked that " education is well advanced, but we are beginning to see that it is like the ancient writing of manuscripts, a slow process with many drawbacks. We labour to perfect the individual, but what we want is, the art of multiplying copies of our best work. Education is waiting for its printing press, which is, scientific propagation." Lord Somerville, in speaking of what breeders have done for sheep, says : " It would seem as if they had chalked out upon a wall a form, perfect in itself, and then had given it existence." Sir John Sebright, so well known as a skilful breeder, says of pigeons, that " he will produce any given feathers in three years, but that he needs six years to obtain head and beak." It is indisputable that vastly more care and thought are devoted to the production of perfect horses, cattle, sheep, and pigeons, than to men and women.

And now let us look for a moment at the question of sterility, which indirectly has a bearing on the subject of sex of offspring. Here, again, we are met by many theories and much conflicting evidence.

Many marriages are fruitless owing to the early deaths of children or to the sterility of the parents, and many theories have been advanced to explain these facts. Consanguinity of parents is with us believed to be the fruitful cause of idiotcy, lunacy, disease, and early death; and when close inter-marriage is continued for generations, these results are believed to be rendered more pronounced. The decay and total extinction of noble families are attributed to this cause. Yet the Jews are most strict in this matter of intermarriage, and they continue to be a remarkably healthy and clever race.

Sterility is relative rather than absolute, unless it

be due to some malformation or infirmity. Perhaps the best evidence to bring out this relative character is that afforded by the intercrossing of races. Some, indeed most, races are fertile "*inter se*," so far as the first cross is concerned. But even with such, where the most perfect hybridity seems to exist, M. Broca assures us—and he has given more attention to this subject than perhaps any other writer—that the mulattoes resulting from the intercrossing of blacks and whites are often sterile with each other, or at least far less fertile than when united to one of the parent races. Sometimes this sterility is not seen until the second or third generation, but it is almost always manifested sooner or later. When the races intercrossed are more remote, a curious phenomenon is presented, called *unilateral* hybridity, *i.e.*, the union of the superior male with the inferior female produces offspring freely, but the converse union is seldom fertile. Again,—and on this point M. Broca is supported by many writers whom he quotes,—instances occur where the union of two races exceedingly remote is nearly always sterile. This is notably the case with the Anglo-Saxons and the indigenous races of Australia. At first sight the foregoing may appear to be remote from the subject of this chapter, but the fact sought to be established is that sterility depends, in part at least, upon great diversity in organization. The special value of this important point in connection with my theory will be fully considered in a subsequent chapter.*

Further as to this relative character of sterility. An individual may possibly have inherited or in

* Chap. VI., p. 144.

some way acquired a constitution with one or more organs so developed—the brain, for example—that a union with one of the opposite sex having a similar excess will surely prove fruitless; but if united to one with a predominant nutritive system, a fair offspring may be the result. Again where the nutritive system greatly prevails in both parents they are almost sure to die childless. These consequences, however, not only result from the excessive development of any one or two organs, but also from every undue divergence from the normal condition.

The late Mr. W. R. Greg believed cerebral development to be a check to fecundity, and he supported this view by citing the lower races of mankind, such as the negroes, and the lower animals. Carey also states the same thing with some modifications. He says: "The general law of life throughout all the classes, order, genera, species, and individuals, may be thus stated:

"The nervous system varies directly as the power to maintain life.

"The degree of fertility varies inversely as the development of the nervous system,—animals with larger brains being always the least, and those with smaller ones the most prolific.

"The power to maintain life and that of procreation antagonize each other—that antagonism tending perpetually towards the establishment of an equilibrium."

Carey also says that the antagonism between the *nervous* and *reproductive* systems is universally recognized. Thus the negroes in America with a low order of intellect are most prolific, as are also the

Irish lower classes. Another fact which might be cited in support of this doctrine is brought forward by Darwin, who finds that the brains of rabbits, and probably of all animals, diminish in quantity under domestication. It is well known that most domestic animals are far more prolific than those in a wild or natural state. Dr. Carpenter informs us that "above *one million* eggs are produced at once by a single cod-fish, whereas in the strong and sagacious shark but few are found." Mr. Carey's opinion is that civilization, in so far as it abolishes drudgery, will cause either physical degeneracy or nervous activity, hence less fertility.

In his work *The True Law of Population*, Dr. Doubleday claims, and with apparent accuracy, that late marriages are far more prolific than early ones. He also holds that scant food stimulates reproduction, while over-feeding leads to sterility. Mr. Elder, in *Questions of the Day*, says:—

"Among the various species of animated beings, we find one invariable and universal fact : The power of reproduction of life is in an inverse ratio to the power of maintaining it. The insects of a day are produced in myriads ; the lower animals, whose span is limited to half-a-dozen years, are reduced and limited to hundreds of offspring ; while the higher grades, that live a score or more of years, are in due proportion less prolific."

Carey cites the sagacious elephant, which produces its young singly, as if its fine brain and nervous system had something to do with this comparative sterility. On the same page he refers to the queen ant of the African termites, that lays *eighty thousand* eggs. Is not the ant quite as "sagacious" as the elephant? It has much the larger brain in proportion to its size. Is there then any exact ratio between *fecundity* and *longevity* in these creatures? Is the elephant *eighty thousand* times longer lived than the ant?

In short, it cannot be said that we have much positive knowledge of the general laws of sterility and heredity. But of individual heredity and its causes we may hope to know much more in time, since the main fact of inheritance of qualities by offspring from their parents or remoter ancestors is patent to every-day observation. The subject of hereditary transmission may best be briefly set forth as it appears in Combe's *Constitution of Man :*—

" At the first aspect of the question, three principles present themselves to our consideration. Either the constitution, size, and configuration of brain which the parents themselves inherit are transmitted absolutely, so that the children, sex following sex, are exact copies, without variation, of one parent or the other; or, *secondly*, the inherent qualities of the father and mother combine, and are transmitted in a modified form to the offspring; or, *thirdly*, the qualities of the children are determined jointly by the constitution of the stock and by the faculties which predominate in power and activity in the parents at the particular time when the organic existence of each child commences.

"Experience shows that the *first* cannot be the law, for, as often mentioned, a real law of Nature admits of no exceptions, and it is well established that the minds of children are not exact copies of those of

the parents, sex following sex. Neither can the
second be the law, because it is equally certain that
the minds of children, although sometimes, are not
always exactly blended reproductions of the father
and mother. If this law prevailed, no child would
be a copy of either parent, but a compound of the
two, invariably, and all the children would be
exactly alike, sex alone excepted. Experience
shows that this is not the fact, but points to the
third idea, that the mental character of each child is
determined by the particular qualities of the parent
stock, combined with those which predominated in
the parents when its existence commenced."

This last must be accepted as substantially the
true law of transmission, by all who admit that there
is any law. I am therefore constrained to adopt the
following :

I. That each parent, at the decisive moment,
transmits of every faculty, organ, and function, just
in proportion to its actual power and activity.

II. That these contributions, in varying proportions at each conception, combine, some faculties
neutralizing each other through excess or other
causes yet unknown, while other qualities, through
felicitous combinations, are intensified or increased,
thus producing a ceaseless variety of human beings.
The law of chemical affinities may serve in a measure to illustrate the idea ; as also may the kaleido-
scope, which by the addition of some fragment,
or a movement which changes but slightly the
relative position of the pieces, produces combinations and figures which it would be impossible,
almost, to foresee ; or the working and tempering

of metals will explain it, where a variation of a few degrees in the heat is sufficient to produce unlooked-for results.

III. That the mother may, and does, exercise a more or less potent modifying influence upon the embryo, throughout the period of gestation.

It thus appears that the result of the conditions producing atavism, reversion, and the inheritance, by children, of the character of both parents, in various degrees of combination, is theoretically quite within our power to calculate ; but actual predetermination with any exactness is rendered difficult, if not almost impossible, by the natural complexity of the determining forces, and by our limited knowledge of their several comparative values. This does not, however, prevent the accurate predetermination of the sex of the offspring of any given marriage.

But what do we really *know* on this subject of inheritance ? Very little indeed. Every one knows that certain idiosyncrasies are frequently transmitted from parent to child, sometimes passing over a generation or two ; but no one can predict with the least degree of certainty how, where, or when these may appear. Most families, however, can point to some special feature or habit which from its constant reappearance has come to be regarded as a family trait.

Statistics show that gout is more often transmitted to sons than to daughters—a disease probably determined largely by persistence in excesses. Supernumerary digits have appeared in certain families for generations,—more frequently, however, in the male line. Consumption is more generally transmitted from mother to daughter, as is also insanity. Idiotcy runs more in the male line. Colour blindness is sometimes transmitted for generations. Some types or features are found to be far more transmissible than others, as a peculiar form of nose, etc. With mules and hinnies, the sire leaves its impress much more perceptibly than does the dam.

Notwithstanding the generally accepted theory among physiologists that the future child partakes and is made up of the qualities most active in the parents at the time of conception; and though this theory appeals so forcibly to common sense, and so many facts corroborative of its accuracy are to be met with on every hand, yet there are points in its application which I find a difficulty in harmonizing.

The parent of twenty, and again at forty-five, or still more at sixty, is sure to have changed greatly, but how is it with the children born at these distant periods? I do not find a corresponding difference in them. Children born of older parents frequently seem to be of a more mature and sober cast of mind, and yet this, I believe, is not a very conspicuous or constant fact. Francis Galton, with his statistics, finds but little, if any, difference between the younger and older members of a family in respect to mental endowment.

Again, that there are many external indications of disposition and constitution few will deny; complexion, colour of hair, eyes, etc., are amongst the best established signs—as will be shown in another chapter. Yet it is an incontrovertible fact that the mother can, by an intelligent control of the imagination and thoughts, during the period of gestation, greatly but indefinitely affect the expression, beauty, colour, moral tone, etc., of the child.

This topic of the mother's imagination it is unnecessary to dwell upon. Nearly every one has some faith in its influence, and most persons can bear witness upon the point, merely relating what has come within the range of their own observation. It might make mothers blush if they knew how many a tale is written so plainly on the faces of their children, that the whole world may read what they had fondly hoped would ever remain a profound secret. Where you find a family of children strongly resembling their father, it is unmistakable evidence that he was almost constantly in the mother's mind during the period of gestation.

These facts, though not numerous, still all point clearly enough to the general law of heredity, and since the most minute peculiarities of parts are transmitted, such as the power of moving the ears or twitching the scalp, a particular curl of the lip, the form of nose, length of chin, etc., it would appear that much is to be said in favour of the hypothesis of Pangenesis (entire begettal), which may be stated in Darwin's own words as follows:—

" According to this hypothesis every unit or cell of the body throws off gemmules or undeveloped atoms, which are transmitted to the offspring of both sexes, and are multiplied by self-division. They may remain undeveloped during the early years of life, or during successive generations; and their development into units or cells, like those from which they were derived, depends on their affinity for and union with other units or cells previously developed in the due order of growth."

This would also explain the above-mentioned belief that the most dominant or active qualities in the parent at the time of conception are likely to be impressed upon the child, for the activity and potency of the gemmules will vary with the degree of activity of the parent cells. Thus, to take an example, the children of drunkards are frequently deficient in brain, if not idiotic or lunatic,—a fact easily explicable if we remember that the brain of the parent was at conception probably deadened and disordered, and that consequently the gemmules which it then threw off were also dull or deficient.

As to the share of both parents in the production of the temperament, features, character, etc., of the child, Professor Huxley's apt illustration in *Science and Culture* may be quoted as the teaching of science to-day:—" It is conceivable, and indeed probable, that every part of the adult contains molecules derived both from the male and from the female parent, and that, regarded as a mass of molecules, the entire organism may be compared to a web of which the warp is derived from the female and the woof from the male "—which illustrates what he has just said, that " impregnation consists in the fusion of the substance of another cell, the male germ, with the ovum, and that the structural components of the body of the embryo are all derived from the coalesced male and female germs."

Again, Papillon says : " According to Mr. Bernard Moulin, children are the living photographs of their parents as they were at the moment of conception :

that parents transmit to their children the tastes and aptitudes of their own physical and mental states the exercise of which were then at their maximum ; " and he cites numerous cases in proof.

As an example of the profound influence exerted by inherited tendencies in the formation of character, the following quotation may be made from Mr. Galton's *Hereditary Genius :* " The gemmules whence every cell of every organism is developed are supposed, in the theory of Pangenesis, to be derived from two causes : the one, unchanged inheritance ; the other, changed inheritance. Mr. Darwin, in his work, *The Variation of Animals and Plants under Domestication,* shows very clearly that individual variation is a somewhat more important feature than we might have expected. It becomes an interesting inquiry to determine how much of a person's constitution is due, on an average, to the unchanged gifts of a remote ancestry, and how much to the accumulation of individual variations. The doctrine of Pangenesis gives excellent materials for mathematical formulæ, the constants of which might be supplied through averages of facts, like those contained in my tables, if they were prepared for the purpose.

" The immediate consequence of the theory of Pangenesis is somewhat startling. It appears to show that a man is wholly built up of his own and ancestral peculiarities, and only in an infinitesimal degree of characteristics handed down in an unchanged form, from extremely ancient times. It would follow that under a prolonged term of constant conditions, it would matter little or nothing what

were the characteristics of the early progenitors of
a race, the type being supposed constant ; for the
progeny would invariably be moulded by those of its
recent ancestry.

"The reason for what I have just stated is easily to
be comprehended, if easy, though improbable, figures
be employed in illustration. Suppose, for the sake
merely of a very simple numerical example, that
a child acquired one-tenth of his nature from individual variation, and inherited the remaining nine-
tenths from his parents. It follows that his two
parents would have handed down only nine-tenths
of nine-tenths, or $\frac{81}{100}$, from his grandparents, $\frac{729}{1000}$
from his great-grandparents, and so on ; the nume-
rator of the fraction increasing in each successive
step, less rapidly than the denominator, until we
arrive at a vanishing value of the fraction.

" The part inherited by this child in an unchanged
form, from all his ancestors above the fiftieth degree,
would be only one five-thousandth of his whole
nature."

It is then no longer possible in the light of modern
knowledge of physiological processes to hold either
the theory of the spermatic or ovularian origin of sex,
for we know that *both elements are concerned* in the
production of offspring, and thus certainly also in the
casting of sex, which is a comparatively late stage or
minute differentiation of the developed ovum. Any
sufficient theory of sex must therefore in future as its
very fundamental principle reckon with the co-opera-
tion of the two parents as factors in the problem, and
this at once and *ipso facto* puts out of court and
demolishes the greater number of the older theories.

Those which are not self-condemned on this principle
are, chiefly, the theories of comparative vigour and
nutrition, genital vigour, and the theory of Doctor
Sixt. The latter we have shown will not bear investigation; the two former, I maintain, are partially
true, inasmuch as they are factors, though not the
only ones, or even the most important, in the determination of sex. They are subsidiary to the true
theory, which takes account of them, but of much more
also. They give us the truth, but not the whole truth
by far; rather, fitful glimpses of the larger generalisa-
tion: and they, taken alone, are wholly incapable of
solving the mysterious results of statistics, whereas
the true theory explains them fully, and even enables
us to forecast them in many cases.

The truth is that mankind has never investigated
the subject, but strangely neglected what might be
positively ascertained with comparative ease. If the
laws of heredity were as well known as they might and
should be, the knowledge of them would greatly conduce to health and length of days, and to the transmission to our posterity of the higher and better elements
of our nature. As matters now stand, the death rate
is very much higher than it should be. The tendency
of all the foregoing is to show that heredity is sufficiently certain and minute to throw a weighty responsibility upon parents in respect of the passions,
predispositions, and tempers transmitted to their
children; and this is an essential part of the theory
of sex which I am about to advance. In this connection the following assertion of the Institute of
Heredity of Boston, in one of its publications, acquires
new force: " We must insist upon the paramount and

supreme right of every human being to be conceived and born in good sound, moral, intellectual, and physical health, and hold all parents to a strict accountability in this matter." It is an established fact that the dispositions, peculiarities, and even temporary condition of the parents tend to impress themselves upon the offspring; it remains for us to examine how far these considerations will guide us in determining sex : their importance in determining character is sufficiently obvious.

And yet, in spite of all this, and knowing our responsibilities to be great, we still take little or no thought for future generations ; and so completely is the propagation of the race left without direction or control, that the word "stirpiculture" is almost unknown as applied to the human race. That this is a great evil none can deny ; and the verdict of those who take the trouble to examine is a terrible one. Mr. Galton speaks strongly on the point, and Mr. W. R. Greg, one of the most incisive and acute writers on social economy, gives his essay on this subject the mournful title of *The Non-survival of the Fittest*. Mr. Greg points out that the constitution, moral and physical, of the two extreme classes of society are the least vigorous and perfect. He says : "The *physique* of the rich is injured by indulgence and excess, that of the poor by privation and want. The *morale* of the former has never been duly called forth by the necessity for exertion and self-denial ; that of the latter has never been adequately cultivated by training and instruction. . . . Yet these two classes are precisely those which are preponderatingly the fathers of the coming generation. Both marry as early as they please, and have as many children as they please—the rich because it is in their power, the poor because they have no motive for abstinence : and scanty food and hard circumstances do not oppose, but rather encourage, procreation. . . It is the middle classes, those who form the energetic, reliable, improving element in the population, those who wish to rise and do not choose to sink,—it is these who abstain from marriage or postpone it."

And Mr. Galton, in his *Hereditary Genius*, demonstrates how this disastrous tendency is cumulative in its operation. He shows that with two classes, one marrying at twenty-two and the other at thirty-three, the first class would in less than a century be twice as numerous, and in two centuries six times as numerous, as the second. "There is," he says, " a steady check in an old civilization on the fertility of the abler classes ; the improvident and unambitious are those who chiefly keep up the breed. So the race gradually degenerates, becoming with each generation less fitted for a high civilization."

Darwin also notices (*Descent of Man*) how we " do our utmost to check the process of elimination by building asylums, hospitals, etc., and by preserving feeble lives which nature would have cut off." Further than this, we restore habitual drunkards and habitual criminals to society with full liberty to perpetuate their vicious tendencies, idiotcy, insanity, etc., without any motive to operate as a check upon early marriage or large families, and, says Darwin again, " it is surprising how soon a want of care, or care wrongly directed, leads to the degeneration of a domestic race ; but excepting in the case of man him-

self, hardly any one is so ignorant as to allow his
worst animals to breed."

We have, moreover, no escape from the application
of this rule to man ; for although we can as yet point
to no example, to prove the beneficial results upon a
nation of judicious care in the matter of stirpiculture,
unless some of the states of ancient Greece be
considered to furnish such evidence, we find in
history plenty of cases of the results of the " care
wrongly applied " which Darwin places by the side of
" want of care " as an agent powerful for mischief in
bringing about the deterioration of the race. Perhaps
no more striking and terrible instance of "care
wrongly directed " can be cited than that of the
Spanish nation under the Inquisition, which is thus
given by Mr. Galton (*Hereditary Genius*):—" The
Church having first captured all the gentle natures,
and condemned them to celibacy, made another
sweep of her huge nets to catch those who were the
most fearless, truth-seeking, and intelligent in their
modes of thought, and therefore the most suitable
parents of a high civilization, and put a strong check,
if not a direct stop, to their progeny. Those she
reserved to breed the generations of the future were
the servile, the indifferent, and again the stupid. . . .
The Spanish nation was drained of freethinkers at
the rate of 1,000 persons annually for the three
centuries between 1471 and 1781. . . It is impossible
that any nation could stand a policy like this without
paying a heavy penalty in the deterioration of its
breed, as has notably been the result in the formation
of the superstitious, unintelligent Spanish race of the
present day."

I have shown, then, how social causes now in operation on a large scale are from "want of care," hindering the perpetuation of the best specimens of the race, and encouraging, or at best permitting unchecked, the unlimited multiplication of the thriftless, the ignorant, and the least able sections of society; and we have seen in history the result of this process of elimination of the best, and continual increase of the worst specimens of the race upon the character of the nation. In other words, we have seen that what Darwin said of a domestic race of animals is equally true in its application to the development of human society, viz., that want of care, or "care wrongly directed," soon leads to deterioration. How great the deterioration now in process is likely to become, if not checked, the example of Spain may afford us some faint conception; but the final result will be infinitely more deplorable, inasmuch as the causes now in operation are more potent, more comprehensive, and more universal. The Inquisition could only prevent a few hundreds of the most liberal and intellectual from handing down to posterity the priceless inheritance of their noble characteristics; but the tendencies of modern society, which bring about the non-survival of the fittest, operate on the whole nation with more or less force, preventing the transmission, not of one or two eminent virtues, but of all the better qualities of the human character; and at the same time they produce an indefinite multiplication of the most degraded natures, *i.e.*, of those having the least self-respect and self-control, with the lowest aims and the strongest passions; in short, emphatically the most unfitted for survival, ensuring at the same time the transmission of numberless inherent tendencies to vice and immorality in every form.

Surely evidence such as that above given—and there is much more of similar character—justifies Mr. Greg's title, and we must all deeply regret this effect of our civilization, which we cannot but believe to be thus far very imperfect; and we shall probably all wish well to the *Institute of Heredity* above mentioned, in its endeavour to find some remedy for this continual deterioration of the race, and shall be ready to welcome any reasonable suggestions which seem likely to compass the end in view. Meanwhile, I believe that when my theory of sex is fully understood and put into practice, the whole tendency of such theory and practice will be to check deterioration and produce a progressively superior progeny in each family; and hence it will tend to promote the elevation of the race. It will, in fact, be the introduction of an elementary kind of stirpiculture, which, notwithstanding the general and deplorable aversion to the subject, all must admit—at least in theory—to be a social necessity, if society is to be saved from continuous deterioration.

Finally, I may summarise the results of this chapter as follows: If parents desire intelligent, moral, healthy, long-lived children, they must not forget that the most promising means yet discovered for gaining that end is to cultivate corresponding qualities in themselves. They must educate their children, withal, to make suitable espousals—they must teach much that is ignored in the present day; and they should remember that instances in which model parents fail to have exemplary offspring are, on the whole, rare. And, further, that the future character and physique of the nation depend chiefly upon the classes of society, and the character and constitution, the conduct and health, of those individuals by whom the race is principally propagated.

CHAPTER VI.

"SUPERIORITY" THE CONTROLLING PRINCIPLE OF SEX.

The control of sex within the power of the "superior" parent—
Various elements involved in the term "superiority"—Mental and
moral qualities not necessarily included—Cerebral development an
important factor—Special influence of the nervous system—Plants
receiving an excess of heat are masculine, whilst more light produces
the feminine—The spinal cord in man and animals—The sun the
source of magnetism—Electricity always circulating in our bodies—
The nervous system the conductor of animal electricity or nervous
force—The electric condition of the human body modified by health
and disease—Also by the state of atmospheric electricity—Electricity
disappears with life and animal heat—May not the nervous force, then,
be convertible with electricity and magnetism?—Positive and negative
electricity or polarity found to exist in the human body—A glimpse
of the great law of sexual equilibrium—The universe in equilibrium—
No theory of sex admissible without some system of balance—This
discovered, the law of sex is revealed—Definition of this law—The
parent endowed with the greater amount of vital force determines the
sex of the future child—An equality of the sexes is thus ensured—
Influence of the superior germ on the sex of offspring—The parental
likeness instantaneously produced—The germs of human life dependent upon the physical and mental state of the parent—The
reproductive system influenced by trouble, suspense, grief, or disease—
The superior germ turns the scale of sex, producing its opposite.

WE have now arrived at the direct consideration
of the law which determines sex, and I will at
the outset state that *sex is determined by what I
shall designate as the "superior parent"; also that
the "superior" parent produces the opposite sex.*

I start from the position taken up at the close of
Chapter III., that the sexes are equal in the aggregate—male equalling female as true physiological
equivalents. But there are many grades of *indi-*

ridual differences and deficiencies in both which will often produce inequalities of sex in single families, just as surely as the aggregate equality of the sexes will produce a corresponding equality in the relative aggregate numbers of male and female offspring.

These individual differences, then, between parents will occasion in one of them what I, for want of a more appropriate term, shall call *Superiority*, and will consequently determine the sex of the offspring. This chapter, therefore, will be devoted to the consideration of what constitutes this "superiority." My difficulty in selecting a suitable designation for the controlling power in each case will be understood, when I state that among the numerous factors to be taken into account are the temperament, activity, energy, will, intellect, features, colour, physique, bodily health, nutrition, etc. The term "Superiority" will be used, but it must be understood to include all these, and to be susceptible of variation through temporary circumstances,—state of health, age, and the like; while the term "*Inferiority*" will signify the opposite. The temperament and energy will be the more permanent elements of "superiority"; but the invidious meaning of the word in ordinary social phraseology must be *altogether eliminated*. Let it be distinctly understood that "*superiority*" *here means everything that tends to increase functional energy* in any part of the system, and thus to influence the determination of sex; and it does not follow that this kind of "superiority" must in all cases coincide with superiority in the general sense of the term—by which we imply superior mental and moral qualities. It may and often does happen that the two kinds of superiority are found together, but the coincidence is not necessary, and they are often, in fact, separated.

Let me now explain in what this "superiority" consists. A true standard should rest upon a solid, scientific basis, and not be left to fanciful conceit. What, after all, do the numberless ideals spring from, except the diverse manifestations of life? And what is more characteristic of life than *feeling* and *activity?* Do not these constitute its essence? Activity is life: stagnation is death. *Life*, then, with all its phases must be the basis upon which to estimate "superiority." However much we may differ concerning what life really is, there can be no disputing over the fact that the brain is its seat of power. The nervous system, if it be not life itself, comes the nearest to its essence of anything we can hope to *see* in this world. The proportion of brain to the weight of the individual is a good general index of capability, in man no less than in the lower animals.

Cerebral development is the key to "*superiority*," as understood in this work. That little mass of pulpy grey and white matter which fills the cranium is the absolute ruler of every organ and function in the entire frame. There are two or three conditions of brain, however, to be taken into account in forming our estimate. *Quantity* will frequently lead us astray, unless we have a suitable regard to its *quality* and *activity:* thoroughly appreciating these points, it is impossible to err in judging people.

" *Superiority*," then—which must be understood as a fuller and higher development of organization—is what determines sex. How it accomplishes this, I will now proceed to explain.

As *Life* is the basis for calculating " superiority," and as the *nervous system* is its apparent abode, it seems proper to scrutinize its manifestations, through this medium, so far as necessary to form some approximate conception of its nature. Let us, therefore, analyze a little and see, if not what life itself is, at least what it is *like*. We see the earth teeming with living matter—animal and vegetable—in such never-ending variety, that it almost seems to spring spontaneously into being. Hardly any portion of the earth is entirely destitute of this profusion, save the frigid zones; and why should they be? To what different conditions are those remote sections of our globe subject? Manifestly, to nothing but a comparative deprivation of *light* and *heat*—of the *sun*, in short. People differ as to the origin and essence of life, but none deny that animal, as well as vegetable life, tends to gradual extinction if deprived of solar influences, which of itself is a most suggestive fact. Botanists tell us, as elsewhere mentioned, that if plants receive more heat than light their gender is masculine, and if more of light, feminine. This may be strictly true, and yet a thousand modifications of the same principle may intervene before reaching man.

In the days of Aristotle and Galen, the phenomena connected with the electric fish were observed and carefully recorded. Scientific men tell us that in the *torpedo*, or *electric ray*, and *gymnotus*, or *electric eel*, the development of electricity does not take place in all parts of the body, but is confined to a peculiar expansion of the nervous system, called the electrical organs. The nerves constituting these organs are of great size : those of the first named consist of three principal trunks, and spring from the cerebro-spinal system ; while the nerves corresponding to the electrical organs of the eel are derived from the spinal cord alone. The effects produced are also similar to those artificially generated, they emit electric shocks. According to Dubois Reymond, there is always present in the spinal cord of all creatures a current of animal electricity which naturally varies with the state of health. In these fish the nervous force is converted into electricity ; in man it is not clearly shown whether this is changed directly into animal electricity or not, but it is obviously not a very material point either way for our present purpose.

Galvani, in 1786, from experiments upon frogs and other creatures, was led to declare that they have a special independent electricity, deserving the name of *animal* electricity. Dr. Ashburner, after a long course of experiments, says that during sleep one is under the influence of attractive magnetic forces, and of repellent forces while awake. He believes the sun to be the source of all magnetism. Dr. J. C. Jackson, who has for many years tested the efficacy of "sun-baths" for the sick, asserts that the more one lives in the sunshine the more vigorous will be his brain.

Prof. Matteucci has unquestionably demonstrated

that currents of electricity are always circulating in the animal frame: that positive electricity is constantly circulating from the interior to the exterior of a muscle, and that muscular contractions are developed in the animal machine by a molecular change of some kind in the nervous centres (*i.e.*, the ganglia of one kind or another), which is propagated along the nerves till it reaches and stimulates the muscles.

M. Béclard is an authority in neurology, and his opinions should not be despised. They are specially to the point. He is led to believe that the nervous system is the elaborator and conductor of an imponderable agent similar to electricity or magnetism, and that by it all the phenomena of innervation can be satisfactorily explained. He instances the relation between the benumbing shock of the electric fish and galvanic phenomena on the one hand, and ordinary nervous action on the other, the practicability of causing galvanic phenomena by the nerves and muscles alone; the possibility of producing muscular contraction, of keeping up the process of digestion, or of continuing the respiratory function of the lung after the nerve has been divided, by connecting the divided portion with the poles of a battery, and substituting electricity for innervation; the existence of a nervous atmosphere acting at a distance around the nerves and muscles, and between the ends of severed nerves. These are phenomena of innervation which undoubtedly resemble those of electro-magnetism, but on the other hand it is an ascertained fact that the propagation of nervous influence is very slow as compared with the rapidity of electrical forces in traversing conducting media.

The comparatively slow rate at which nervous force travels—about 100 feet per second—is noteworthy, yet it should no more cause us to question its being allied in nature with electricity, than the colour and coagulation of blood and milk, respectively, should cause us to doubt that they have water for a common basis. According to the experiments of Helmholtz and Schleske with the chronograph, the velocity of the nervous force in man is 97·5 feet per second, which is not more rapid than the speed of the greyhound. By means of the noematochrograph it is found that the brain is *one-twenty-fifth* of a second in recognizing an impression, and *one-twenty-eighth* in telegraphing that an impression has been received; also that *one-twentieth* of a second is required to distinguish and signal the difference between two colours.

Hemmer, in 1790, and Gardini, in 1793, experimented extensively to ascertain the general electric condition of the human body. They concluded that there is a free electricity in our bodies, sometimes positive, and at others negative, variously modified by conditions of health and disease. Hemmer found positive electricity in himself 1,252 times, negative 771, and on 399 occasions neither. He concluded that positive electricity is natural to the body, modified by physical and mental exertion. Sfösten in 1800, Ahrens and Pfaff in 1812, and Nasse in 1834, investigated the same point, and severally concluded (and let their conclusions be especially borne in mind) that free electricity is more manifest in the enthusiastic and excitable than in the cold and phlegmatic ; and furthermore, that we have more of it at evening. Nasse believed that there is positive electricity in all, and at all times.

The animal system in health, as well as in disease, would seem to be variously influenced by the conditions and variations of *Atmospheric Electricity*. The facts and experiments on this subject have been so limited and unsatisfactory that most people, including men of science, have very properly suspended their judgment in regard to the matter. But so much is now known that few will dare dispute that all—more especially those of a *nervous* temperament—are appreciably and traceably influenced by the variations in the quantity and quality of the free electricity in the surrounding air.

Drs. Beard and Rockwell have given so much attention to medical and surgical electricity as to be recognized authorities, and they have published much valuable information. They say that currents of electricity have been found not only in the muscles and nerves, but also in the central, and even in the sympathetic nervous system. Electric currents, too, are found in the skin, spleen, reproductive organs, kidneys, liver, and lungs, as well as in the nerves and muscles. The sheaths of muscles, fasciae, and sinews do not contain electricity. As life leaves the body, the indications of electricity decrease and disappear with the animal heat, and electric or galvanic currents well-nigh cease to produce any effect upon a corpse. They most sensibly conclude upon the point thus: "What is the relation of electricity and magnetism to life? If light and heat, and motion, and electricity, are mutually convertible, may not the *nervous* force also be convertible with electricity and magnetism?"

I think that enough has been adduced to convince
the reader that either an electric, magnetic, odyllic,
or some kindred force, however designated, pervades
every living human frame. To show the application
of this fact to our subject, I would call attention to
the most striking characteristic of these mysterious
properties of matter : It is well known that when any
magnetic substance is subjected to the influence of
electricity, instead of diffusing itself equally throughout the mass, this subtle fluid assumes what is termed
a positive and negative position, or *polarity*—the
distinctive principle to which attention is now called.
We know several of the peculiarities of this phenomenon, but *why* such a condition should instantly
result, none can explain. The North Pole of the
earth is positive, and the South negative, yet the
north pole of every balanced needle is negative, and
the south positive ; which in part illustrates this perplexing enigma of *polarity*. If a rod of metal be
magnetized, one half is found to be positive, and the
other negative. If this rod be cut into twenty pieces,
each one of them becomes a magnet, as perfectly
polarized as was the first.

Reichenbach and Ashburner find that the human
body possesses this same positive and negative
property. They find that in human beings, males
and females are polarized in precisely the same way,
the general polarities being in the right and left
sides, in the hands and feet, besides countless minor
points. They find the whole left side of the human
body positively, and the right side negatively charged,
irrespective of sex.

Quite in harmony with the foregoing is the fact

that an injury on one side of the head causes paralysis of the *opposite* side of the body. Physiologists account for this by alleging that there is a *decussation*, or crossing, of the nerves, near the base of the brain; in fact, they seem to consider it as almost a self-evident proposition. Maudsley, however, very pertinently asks: "Where does the decussation of the fibres for sensation take place?" and in a way which shows that he has serious doubts on the point. I would venture to suggest that the true explanation may yet be proved to be something very different from the one now accepted—that it is but a phase of polarity.

Though it may or may not be due to decussation of the nerve fibres, I think that no one can doubt that this "crossed action" does really exist in the human frame, and to my mind it has a most significant import for our subject, inasmuch as it gives us what I believe to be a glimpse of the great law of sexual equilibrium. Friction evolves the two electricities, and it is now known that the quantity of *negative* electricity thus generated is always precisely equal to the *positive*. Again in the matter of gravitation: we see revolving bodies apparently subject to two opposing forces, of which we know just enough to call them *centrifugal* and *centripetal*—the one tending from and the other towards the centre. There seem in fact to be two opposing or counter-balancing principles throughout Nature which thus keep the universe in stable equilibrium. Cause and effect are everywhere acting and reacting with perfect uniformity and precision.

To devise any theory of sex without being able to

accompany it with some element of balance for its foundation, is mere speculation, for it is precisely the prompt action of this self-adjusting principle which has won the admiration of all who have attempted its discovery. To rightly discern this power of equalization is almost to know the law of sex itself: it is more than the *key*, it is the open door thereto.

The foregoing will have prepared the reader for the statement of what I claim to be the law of sex. I define the law of sexual equilibrium as *that ever visible principle, of all orders of animal life possessing a division of function, by which the one having the preponderance of vital endowment turns the scale in casting the sex and bestows upon the embryo the gender of the other weaker parent; the offspring being thus endued, approximately, with the force of the superior, an abiding equality of the sexes is ensured to each species.*

Our first consideration must now be concerning the nature of the life-force which determines "*superiority.*" The analogy between photography and reproduction will perhaps serve as the most suitable comparison presentable.

When a photograph is taken, the actinic rays of light execute the masterpiece, and do so instantly; it is the only known agent capable of such a feat. In the generative act, the exact likeness of each participant may be said to be struck off with the rapidity of lightning. Like other electrical phenomena, it is practically *instantaneous :* is this nothing more than a coincidence? Have not these forces a common origin?

A single mustard-seed is large enough to contain

thousands of the seminal granules, which are only $\frac{1}{5000}$ of an inch in diameter, while the ovisacs, which are the vesicles of the future ova, and are termed Graafian vesicles, are from the size of a pin's head to that of a pea, and it may seem absurd to suppose that a man in miniature—or his very essence—is contained in a few such particles, and yet results show that something equivalent to this is the fact. Nor is it any more strange than the fact that a tiny acorn is a germ which, under favourable conditions, will produce the giant oak.

All vegetable germs, or seeds, are as much of a marvel as the acorn, while those of animals are, if possible, still more wonderful, for their seed is not made in profusion, periodically, capable of being transported and preserved indefinitely, but is produced speedily in the breeding season when occasion requires, and as quickly loses its virtue if neglected. After all this is considered, it need not be surprising that human seed varies with our daily fluctuating conditions. A tree yields substantially the same fruit year by year, but in living beings the sex and character of the offspring depend upon the physical and mental state of the parent at the hour of its formation. This is a most essential point to be borne in mind. Our bodies are constantly undergoing change, as also are our miniatures—our life germs. We must disabuse our minds of the idea that the spermatozoa and ova are imbued with any abiding character : organic *change* is the incessant order of life.

Insects undergo complete metamorphosis sometimes more than once in their lives, and the higher forms of animal life—man included—are continually undergoing a process of local death and regeneration. Our bodies are partially altered in substance by every meal we eat, and so continuously does this change go on that it is said there is a complete change of every molecule of the body in a period of seven years. A similar term is found in physiology to denote periods of change in the constitution. If this change be thus radical in seven years, it must be continually operating; and as there are also many causes of temporary constitutional change, such as trouble, excitement, suspense, grief, and disease, we shall have no difficulty in comprehending that these differences and changes will appear in a more or less pronounced manner in offspring conceived while the parents are under their influence.

There exists the most active sympathy between the cerebral and reproductive systems. Dr. Clarke, in his famous "*Sex in Education*," bases much of his argument on this fact. Both sexes are alike susceptible to this response of the reproductive organs to every agitated mental state, or injury of the nervous system; even in the criminal, dying on the gallows, we have most unmistakable proofs of these organs being thrown into the very highest degree of convulsive excitement by the mechanical injury done to the spinal cord.

Fortunately, however, the evidence of the union of these two systems is not left to the *ipse dixit* of the author. Drs. Beard and Rockwell give the following as the result of their professional experiences:

" The *generative organs* are frequently corrected in

their functions by the application of electricity to some part of the body. These organs are so intimately connected with the vital parts of the system by means of the gauglionic nerves that they must necessarily share in all the good or evil influences affecting the spine or the body generally."

Dr. J. S. Jewell, Professor of Nervous and Mental Diseases in the Chicago Medical College, and Editor of the *Chicago Journal*, writes of these diseases, in that periodical, under the heading of "The Relation of the Nervous System to the Uterus," stating facts which it will be of interest here to note. He, at the close, says:

"But to sum up the sources of uterine nervous supply. They have a double origin:

"(1) *Cerebro-spinal.* . . . . .

"(2) *Sympathetic nerves.* These are derived from:

"A, The bifurcation of the abdominal aortic plexus, at the level of the sacro-vertebral angle;

"B, The inferior mesenteric plexus;

"C, The renal plexus;

"D, The lumbar ganglia;

"E, The sacral ganglia.

"These nerves taken altogether contribute to form two plexuses, the ovarian and the hypogastric, that are distributed as already described. . . . . .

"From this outline of the anatomy of the nerves of the uterus, you may gather: That the uterus is bountifully supplied by nerves, both from the sympathetic and cerebro-spinal nervous systems, and that its nervous connections with other organs are both numerous and important.

"That there are probably in the lower part of the

spinal cord one or more nervous centres from whence the nerves proceed that more or less supply the uterus.

" That in the human being there is some anatomical ground for believing the genito-spinal centre to be at or near the point of origin in the spinal cord of the fourth sacral nerve." . . . . . .

For a long time it has been regarded as probable that the spinal cord has a direct influence on the organs of generation. But it was Professor Budge of Greifswald who probably made the first definite experimental observations in regard to the influence of the lower part of the spinal cord on the motions of the generative organs.

Dr. Jewell's concluding remarks on this point are as follows : " These earlier experiments of Budge have since been controverted to a certain extent by Kupresson, but they have been reaffirmed by Budge, so that there can be but little if any doubt that a genito-spinal centre does exist in the lower part of the cord, which may be regarded as the immediate spinal nervous apparatus of the uterus. But why should we hesitate to admit that such a centre exists in the cord, even if the experimental proofs of its existence are not perfect, when we readily admit that respiratory, cardiac, vesical, anal, and other centres exist in some part of the nervous system ? "

I think, however, that the existence of a genito-spinal centre is now clearly enough established, and if so, it furnishes the missing link and enables me to close this chapter by briefly summing up my theory :—

The reproductive germs in both male and female, whether prepared immediately or days before their contact, are modified by physical and mental conditions up to the very last moment, as are all our secretions. The *life-force*—by whatsoever name it may be designated—which acts when the germs unite is governed by general laws and displays striking analogies to the magnetic forces in Nature. The superior germ rules—as "superiority" is sure to do, wherever it be placed—and, out-balancing the other, turns the scale of sex, but, subject to that unvarying magnetic law of *opposites*—which I conceive to embody the principle of universal equilibrium—produces its opposite. Hence, if the mother be more highly endowed, or even but temporarily more favourably conditioned than the father, the offspring will be of the male sex, and endued with more of the mother than of the father. The converse of this of course will obtain, and a "superior" father, on the same principle, will beget a daughter. *The theory may appropriately be designated as that of* SUPERIOR OPPOSITES.

I am aware that nothing conclusively definite is here stated as to the real nature of "superiority." Nothing positive is, however, known as to what constitutes life-force, energy, etc.,—what it is, in short, which makes the difference between one individual temperament and another ; and while science is thus in the dark, no accurate definition can be given. But enough has been said to give the reader a clear idea of what is meant by "*superiority*," and what it accomplishes, though what causes it, and what is its precise nature, must remain for the present obscure.

But how to determine accurately which *is* the "superior" parent, and the way to become "superior"

at will; how I was led to the discovery, and the
irrefutable facts in history, together with data of a
scientific, genealogical, and statistical nature, which
render my position impregnable; the light that this
discovery throws upon old theories; the practical
value of this knowledge in every-day life, and how
it enables us to solve numerous moral, social, and
political problems, it will be the purpose of the
succeeding chapters to explain.

CHAPTER VII.

WHAT INDICATES AND DETERMINES " SUPERIORITY."

The external, visible marks of "superiority"—At least three or four
qualities necessary to constitute "superiority"—"Superiority" as
indicated by the temperament, complexion, will, etc.—Characteristics
of the nervous temperament—The nervous-bilious temperament defined—Advantages and disadvantages of the sanguine temperament—
The lymphatic temperament—Persons of this class decidedly "inferior"
—Activity an indication of "superiority"—A well-developed nervous
organization "superior" to all—Complexion—Dark complexions
generally "superior"—The colour of the hair, eyes, and skin in relation to human temperaments—Fair persons should marry those of
red or dark complexions—Will or decision—The result of the combined influence of several organs—Napoleon a remarkable example—
Will-power favourable to "superiority"—The muscular system—The
nutritive system low in the scale—Weak digestion often accompanies
an active brain—Personal marks of "superiority"—The head and its
indications—The forehead—The veins—The brow—The eyes—The
nose—The mouth and lips—The teeth—The chin—The face, cheeks,
and neck—The hand—Activity, physical and mental—Spasmodic
and incessant activity—Attempt to analyse activity—Activity of
speech usually favourable—Also activity of walk—Consideration of
the couples figured in the forty portraits at the end of the work—
Race a factor in the determination of sex—Extreme inequality
favourable to sterility—Bearing of this fact on the origin of species—
Its influence recognised by the late Mr. Darwin—The principle of
adjustment elucidated—Individual "superiority" not overruled by
heredity, but accounted for.

I SHALL in this chapter define the external, visible
indications of "*Superiority*." This is a very important part of my theory, and with the ample
explanations given and the illustrations at the end
of the book, no one, I think, will experience any
difficulty in discovering the "*superior*" of any given
couple. There is no single infallible point, or feature,
which will settle the question in every case; man

is such a complex machine that it is impossible to decide without taking into consideration at least three or four of his distinguishing mental and physical traits, both permanent and transient.

As in treating of population we found different factors involved, so now, in dealing with individuals, we find different standards from which to make our estimates of "superiority." In society we are disposed to class men according to their titles, their fortunes, or their general appearance. Nature, however, adopts a different standard from any of these : it must be borne in mind that it is *Nature's* estimate that I am here trying to present ; and I doubt not that the more one observes and judges for himself, especially among his acquaintances, the more convinced he will be that Nature substantially determines the sexes in accordance with the principles I have set forth.

It is manifest, from what has been said in the preceding chapter, that there is no *direct* method of determining the " superiority " of any individual, but " superiority " will assuredly make its existence evident by various characteristic indications on the physical frame or the mental constitution of the individual, and by these we must judge. I propose to consider a few of these in the following order :—

Temperament, as indicated by general appearance and habits.—*Complexion*, as indicative of constitutional differences.—*Will*, as characteristic of mental qualities.—The *reproductive system*.—The *muscular* and *osseous system*.—The *nutritive system*.

Special features.—Head—forehead—veins—brows

—eyes—nose—mouth and lips—teeth—chin—neck, etc.—Activity.

"*Superiority*" as indicated by the *temperaments* is, then, the first topic to be considered.

The possession of any one temperament cannot be adduced as an unfailing indication of real pre-eminence, for the reason that the temperaments are seldom found singly in any individual, but are combined and blended in ever-varying proportions. Besides this, there is perhaps no point upon which authorities differ so strangely as what temperament really is, and in determining to which class any given individual properly belongs. Some arrange mankind under three or four heads, while others make six or eight divisions.

I consider that a particular temperament corre-sponds to a predominance of one or more systems of the body, or to a deficiency of one or more of them. Thus the cerebral, nutritive, circulatory and lymphatic systems may predominate and give rise to special temperaments.

The question is certainly one upon which there is no very complete and settled theory; but it may be admitted that to be of the nervous temperament, for instance, is merely to have the nervous system in the ascendant; and also, as has been said, "a balance of all the temperaments is to be of no special temperament."

Letting this, then, suffice as an explanation of what is meant by "temperament," I will proceed to show the relation which it bears to "superiority." I shall speak of the temperaments under the most common classification of Nervous, Bilious, Sanguine, and Lymphatic; and it will be seen that the descriptions or definitions which will presently be given of each temperament agree fully with the idea of the predominance of one or other of the above four physiological systems in the organisation of the individual. Moreover, I consider that in establishing "superiority," the temperaments take precedence in precisely the order here given; any one with the cerebral system suitably preponderant is "superior" to all other combinations, while one with the lymphatic is inferior to all.

Having found in a Spanish volume, by Cortés, what I consider as satisfactory a description of these four temperaments as I have seen in any language, I translate his delineation rather than attempt an original one; indeed, had they been my own words I might have been charged with drawing upon fancy, merely to suit my purpose. The points which especially indicate "SUPERIORITY" I have printed in capitals, while the unmistakable signs of "*inferiority*" I have put in italics.

First then :—"The nervous temperament characterized by cerebral predominance and known by a HIGH, BROAD HEAD, and a STRONG INCLINATION TO MENTAL LABOUR; GREAT VIVACITY OF FEELING, an ACTIVE IMAGINATION, and a RAPID SUCCESSION OF IDEAS, which seem to emanate spontaneously. The bodies of persons of this temperament are of slender form, sometimes dry; their muscular system is frail, and of more than a normal thinness. Their skin is slightly flushed, generally fine, and VERY SENSITIVE; they have a brilliant eye, but with a *timid expression*. All their functions change easily, and since they are of strong emotions they are much inclined to love. Their SLEEP IS LIGHT and *disturbed by dreams:* their ailments soon become complicated with cerebral symptoms, and the head generally becomes the seat of the disease."

Of our illustrations, A 2 and A 6 give us a pretty good idea of this temperament. The undue expanse of brain is very evident in persons of this class. They are inclined to be dyspeptic and to suffer with sick headache.

As I have already said, there is a never-ending variety of combinations. In the above quotation, the nervous development is so extreme, and so much at the expense of other parts, that it by no means represents my ideal of "superiority." In fact, it hardly equals the following outline of the Bilious— which surely must have an excellent nervous accompaniment to fulfil all the conditions, and I should call it *nervous-bilious,* and on the whole superior to the first named.

Secondly, the Bilious temperament characterized by predominance of the stomach, intestines, liver, and spleen. Those who have these organs largely developed, and more active than other parts, are DARING, OF A FIXED AND PENETRATING EXPRESSION OF EYE, WITH PROMINENT MUSCLES, AND GREAT VIVACITY OF MOVEMENT. They are of medium stature, and have a DARK SKIN, DRY, WARM, and HAIRY, characterized by VEINS THAT STAND OUT IN BOLD RELIEF. Their hair is BLACK, OR DARK CHESTNUT, and falls early.* They digest food easily, and can ENDURE FASTING. These SLEEP LIGHTLY, AND REQUIRE comparatively

* This I have not found confirmed by observation.

LITTLE OF IT. They usually DEVOTE THEMSELVES TO THE ACCOMPLISHMENT OF A SINGLE OBJECT, HAVE GREAT CONSTANCY OF PURPOSE, A POWERFUL IMAGINATION, and are APT TO RIDE HOBBIES. They are inclined to be EGOTISTICAL, SUSPICIOUS, and to EXAGGERATE THEIR FEARS AND WRONGS; they are INDEFATIGABLE IN THEIR ENTER-PRISES, and MOST PERSISTENT, ZEALOUS, passionate, and revengeful. They are especially subject to disorders of the stomach, or intestines, and above all of the liver, which makes them MELANCHOLIC. This temperament has furnished the world with TYRANTS, CONQUERORS, BENEFACTORS TO HUMANITY, and legislators."

The reader will better understand the reasons for characterizing some of these traits " superior " and others " inferior " as we proceed. A 1, A and B 4, A 5, A 9, and B 17, show us modifications of the temperament above described. A 4, has almost too much brain to be mentioned here, while A 9, has scarcely sufficient cerebrum, yet an abundance of intensity and activity: " constancy of purpose " is conspicuous in B 17. B 4, has not much brain, yet she is emphatically of the bilious type.

In looking for this temperament we must not expect to find a corpulent abdomen to contain the " predominant stomach; " the largest eaters have not remarkable abdominal proportions. This temperament eats not to lay up fat, but to meet the daily expenditure: an engine that consumes but little fuel cannot accomplish much. A hearty eater, of this temperament, is not necessarily a glutton.

Third, " The Sanguine temperament, characterized by predominance of the lungs, heart, arteries, and veins. Persons of this class have a broad chest, a large heart, and have easy and ample respiration. Their veins are superficial and well developed; they are tall, with chestnut or *light hair, white* and *rosy complexion, EYES AND COUNTEN-ANCE ANIMATED*: they have considerable natural warmth and *moderate appetite.* They eat and digest with rapidity, and *sleep with ease;* they are, consequently, strong and happy. Their ailments are usually of an inflammatory nature, rapid in their course and terminating favourably.

" If there be a GOOD CEREBRAL DEVELOPMENT, they have an EXCELLENT MEMORY, ACTIVE IMAGINATION, and FONDNESS FOR LABOUR THAT EMPLOYS THE UNDERSTAND-ING, but they are *not given to meditation.* They are benevolent and happy; their *anger* is easily kindled, but it as *speedily vanishes;* all their passions are lively, *inconstant,* and are sure to pre-sent *marked defects* or BRILLIANT QUALITIES. If this temperament be united with the bilious in an individual, he possesses the advantages and dis-advantages of both, modified in a certain degree, however, by the combination."

Probably A 15 and A 16, have more of this sanguine element than any others among my illustra-tions, although A and B 12, are not wanting in the same. I cannot imagine A 15, with anything but a florid or red face. The combinations are infinite: A 8, has a share of the sanguine, almost overpowered by the bilious element.

Fourth, " The Lymphatic temperament, charac-terised by predominance of the cellular tissue and

white fluid, or Lymph. Those of this temperament
have an abundance of white corpuscles, their muscles
being, as it were, infiltrated with these humours, but
not rendered prominent thereby. Their lungs, heart,
and arteries are but poorly developed, and their *blood
is less fluent* than in the other temperaments. They
are subject to *corpulency*, but it is not healthy *fat.
With the least sickness they become extremely thin,*
and any occupation requiring some degree of activity
they find themselves *unsuited for*, and they are *soon
disheartened and depressed ; their fatness vanishes,*
leaving them much emaciated. They generally waste
away only in those parts most exercised. They present a *pallid and swollen appearance :* their *eyes* are
delicate and *expressionless :* their *lips* are *ample ;*
their *hair* is *light* or chestnut, and *lies sleek* and
flattened to the head. They appear to be *unnaturally fat*, and there is *no consistency in their skin or
flesh.* They *eat little and digest but poorly :* they
sleep long and soundly. They are *slow and heavy in
their movements. Pleasures* of any kind, even those
of love, *scarcely arouse them :* they are *wanting in
intelligence :* they have *no memory or penetration :*
they *are lazy, insensible to the charms of the arts and
sciences.* They are *well-nigh passionless.* They have
however some judgment; they are *easily vexed, but
soon become placid again.* They *soon forget wrongs
and insults*, which in fact *make but slight impression
on their minds.* They are good, affable, benevolent,
indifferent to most that is transpiring around them.
No great sacrifices should be anticipated from them.
They are usually quite happy, for they may be said
merely to vegetate."

This description undoubtedly answers to a pretty strongly marked type of the lymphatic, though not one in a hundred probably could be found so dull as here described. The illustration A 18, can hardly be improved upon for our purpose, and B 2, will also come to substantially the same condition, at the same age. B 8 tends in the same direction, but is not such a decided case. A 14, is unmistakably lymphatic, as well as A 17, yet their open eyes show that they are in no immediate danger of being deteriorated by the excess of lymph.

I think, on the whole, that the reader will ere this be disposed to agree with me that *activity* indicates "*superiority;*" that the lymphatic is decidedly the "*inferior*" temperament, and that a well-developed nervous organization is "*superior*" to all others.

Complexion. — Colour has much significance throughout Nature; and it is safe to affirm that dark complexions generally denote "superiority," while the light, or pale, are tokens of "inferiority,"— in the vegetable kingdom plants and trees of the darker hues are usually the more vigorous, and dark-coloured fruits the most delicious.

This rule seems to apply with equal force to the lower animals. Most people act upon it in the purchase of their domestic animals. The following quotation from Dr. Horne gives us an idea of the relation between colour and character as found in the horse, and I feel sure there will be few dissenting voices among those best acquainted with this noble quadruped :—

" Dark chestnut, true bay, and dark brown are the best colours. Horses of these colours I find to

be freer from disease than those of other shades,
and most certainly do I hold them up in the front
rank for their general qualifications—speed, stamina,
docility, size, and beauty of conformation. I know
colour does not make this difference, but the peculiar organization which makes the colour does."

In the human race we find substantially the same
thing. The darker individuals *of any given nationality*, other things being equal, are superior to their
paler faced brethren; but this must not be applied
to the comparison of different nations and races,
whose colour is modified by many complex causes.
The following indications of character founded upon
the colour and texture of the hair, is a quotation
from a book published in Paris, entitled, " Secrets of
Beauty." It contains probably as much truth as will
generally be found in any such sweeping generaliza-
tions. I have italicized those portions of it which
my observations have assured me are generally true.

" *Coarse black hair and dark skin signify great
power of character, with a tendency to sensuality.
Fine black hair and dark skin indicate strength of
character along with purity and goodness. Stiff,
straight black hair and beard indicate a coarse,
strong, rigid, straight-forward character. Fine dark
brown hair signifies the combination of exquisite
sensibilities with great strength of character. Flat,
clinging, straight hair, a melancholy but extremely
constant character.* Harsh, upright hair is the sign
of a reticent and sour spirit, *a stubborn and harsh
character. Coarse red hair indicates powerful animal
passions, together with a corresponding strength of
character.* Auburn hair and a florid countenance

denote the highest order of sentiment, intensity of feeling, and purity of character, with the highest capacity for enjoyment or suffering. Straight, even, smooth, and glossy hair denotes strength, harmony, and evenness of character, hearty affections, a clear head, and superior talents. *Fine, silky, supple hair is the mark of a delicate and sensitive temperament,* and speaks in favour of the mind and character of the owner. *Crisp curly hair indicates a hasty, somewhat impetuous, and rash character. White hair frequently denotes a lymphatic and indolent constitution.*

"Besides all these qualities, we may add that there are chemical properties residing in the colouring matter of the hair tube which undoubtedly have some effect upon the disposition. Thus *red-haired people are notoriously passionate.* Now, red hair is proved by analysis to contain a large amount of sulphur, whilst very black hair is coloured with almost pure carbon. The presence of these matters in the blood points to peculiarities of temperament and feeling which are almost universally associated with them. The very way in which the hair flows is strongly indicative of the ruling passions and inclinations, and a clever person could give a shrewd guess at a man's or woman's disposition by only seeing the backs of their heads."

Dr. Powell, in his " Human Temperaments," points out with considerable precision the colour of hair and eyes, and the complexion by which the temperaments, in all their numerous combinations, may be known. He makes black or dark hair, eyes and skins spring from the bilious element, and the light from

the sanguine. Where a lymphatic subject is found
of sombre type, he claims that the bilious is the
basis, and the sanguine the source of the light
varieties. He asserts that but two of the temperaments have any constitutional basis ; the encephalic,
or nervous, and lymphatic, he holds to be mere
parasites, or adjuncts, the growths of civilization and
wealth. He gives us many of the mental characteristics of what he calls the bilious ; but he *describes*
rather than *defines* this and other temperaments.
He presents one point which tends to confusion, and
if true—as it certainly appears to be—it shows us our
imperfect knowledge of this matter of colour. He
mentions the " *red bilious*," or " *.xanthous*," which
differs in no way from the dark bilious, save in colour.
It will be seen by the quotation from " Secrets of
Beauty," that "coarse *black* hair " and " coarse *red*
hair " are ascribed to persons of the same general
traits. Dr. Pritchard, the ethnologist, found that
the progeny of dark Jews emigrating from Palestine
to North Germany were red. The same result is
known to follow those from Louisiana who settle in
Pennsylvania and Ohio. Dark birds are said to
become decked in white plumage if taken to Siberia.
The change appears to be confined to the dermal
system, whether showing in hair, feathers, or skin.

Dr. Powell calls red a bilious indication, since it
accompanies the same cranial conformation as we
find in persons with dark hair, and both individuals
possess the same characteristics. He also finds that
mercury and other drugs and poisons act in precisely
the same manner upon each of these classes, whereas
widely different results are obtained amongst persons

of other temperaments. My own observation convinces me that he is quite right upon this point.

On the whole I feel justified in declaring that, in a large majority of instances, dark hues have a decided advantage over the light ones, and are " *superior*." Therefore let all those of fair complexion seek to unite with those of red or dark complexions, for the pale, light types are very generally less vigorous and less enduring, and therefore should not be united. Two of a dark type may safely marry, but not two who are fair, especially as the union of the latter is likely to result in delicate offspring.

Will, being one of the visible indications of " *superiority*," must be taken into consideration. It is *not*, however, a very *exact criterion* of the comparative worth of individuals, as I shall endeavour to show. Webster defines will as " the power of choosing," but *decision*, in a single word, best expresses my own conception of it.

The will is not embodied in any single mental faculty, but every act of volition is prompted by the combined influence of several organs; it is the voice of the majority: the faculties most active at any given time determine the action of the will. Some persons are said to have strong wills, and others weak ones, but we seldom meet with an individual who has a will of uniform strength; the will is found strong in some directions and weak in others. Sir Isaac Newton, whose intellectual powers are unquestioned, must be classed as *superior* in almost every sense of the term, and yet he was noted for his indecision, or deficient will-power. This need surprise no one, for Newton's forte was profound thought, abstract reasoning. Coleridge was still more undecided, chiefly owing to his habit of taking opium, which enfeebled his whole mind : he lacked power of concentration,—hence many of his poems are "fragments." He was metaphysical and poetical, and, giving free rein to his imagination, was inclined to sit and dream life away.

A well-balanced mind is indispensable to a consistent will, and, as in everything else, exercise, or cultivation, is essential to a *strong* will. Decision of character is to a great extent a *habit of mind :* one can generally contract or eradicate a habit of this kind by due vigilance and effort. Some persons seldom find themselves unsettled and vacillating, while others are continually arriving at a doubtful point and have to ask advice : they have "no mind of their own." Dr. Henry Maudsley, in "Body and Mind," states that "volition is like memory, a physiological function of the nerve centres, and becomes unconscious and automatic the more completely it is organized by repeated practice. Loss of *will-power* is one of the earliest and most characteristic symptoms of insanity, and is the direct effect of physical disease." In this I concur, save that most frequently it is a slack mental habit and not physical disease which is the cause of deficient will-power.

Holding a responsible position is the best means of cultivating decision of character, or the power of will. To make perfect self-control one's constant aim is also sure to develop this function. Rulers and military officers readily develop a strong will ; much of it is born with them, no doubt, but much is also acquired. Napoleon had a mind which needed but little cultivation to become perfect in will-power, making the most inflexible resolves in an incredibly short space of time.

Activity, though not adding directly to will-power, accelerates *decision*. Of two individuals, otherwise equally endowed, but differing greatly in activity, the one will resolve and execute, while the other is deliberating: but this element will be more specifically discussed further on.

The manifestations of will, then, though not in all cases positive indications of "superiority," are favourable circumstances as far as they go, just as surely as their absence is a token of "inferiority." In this connection the contrast between faces like A 5, A 8, A 9, B 17, and such as A 18, B 7, and B 8, are very striking.

A 7 is the head of a profound philosopher, but we see that, like Newton, he can never become conspicuous for decision of character; he is lost in profound abstractions. A 19 exhibits more determination by far, and yet cannot be called the superior, although possessing more of what is popularly termed *will*. He can be noted for little else than obstinacy, while the philosopher has a brain teeming with thought.

The *reproductive system* is most important in its influence upon the general organisation, as it is inseparably allied with the nervous system in its action. In the lower animals its development in a normal way can be observed and compared with the other systems to better advantage than in mankind.

The nervous system is beyond all dispute the superior, the crowning one of our being. The man in whom it predominates will generally be of intellectual tastes, and pursuits tending to develop his better nature. We can with difficulty imagine that there is anything debased in individuals like A 2, A 3, B 16, B 17, etc.

This reproductive function, depending, as I have said, so much upon the nerve force for its action, gives vigour and tone to the whole frame, and, if kept under proper restraint, is conducive to " superiority." Unfortunately with a majority of men this function is allowed to obtain the ascendant, thus debasing their nature, instead of being kept under restraint, and thus conducing to a higher life.

In A 1 we have a noble type of manhood, and yet a person led by passion,—which shows that intellectuality can exist, for a season at least, linked together with lust. One can hardly look at faces like A 8, A 16, A 19, or B 6, and imagine them to be of deficient or dormant sexuality. Yet I am led to believe that nothing is more difficult than accurately to judge of the passional nature. A 6, A 7, A 18, B 9, and B 17, appear to be less enthralled by amativeness than any others in the list.

The predominant *muscular system* is well illustrated in the thrifty, temperate blacksmith. It is not to be despised, especially when not developed at the expense of brain culture, as very frequently happens among the labouring classes. The *osseous* frame is intimately connected with the muscular system, and an excessive development of the bony structure must surely be at the expense of some other parts; it rarely characterizes one of acknowledged brilliancy. Still, muscular exertion tends to invigorate the whole system, and therefore all athletic exercises, when moderately employed, are valuable in raising the general tone. Persons largely endowed with bone and muscle, however, naturally make their way to some colony to battle with the virgin forests, rather than incline their steps towards the forum, the pulpit, or the office desk.

Almost all the full round faces in this book show muscular development, or flesh; yet there is a great difference in the value or quality of what each possesses; A 8 has good, active flesh, while B 8 has an undesirable kind. A 14 is somewhat of the same class, while A 18 has an inferior frame in every way. The osseous system is generally manifest in the nose; of course bony extremities are sure criteria —as in B 4, B 7, and A 9.

The predominance of the *nutritive system* is indicated by rotundity of features, ease, laziness, corpulency, etc. To compare this system with the reproductive and muscular, place the stall-fed ox in line with one from the yoke, and a Durham bull next; which will be the "superior"? Is it not an ascending scale of good, better, best? Or, again, take the blooded stallion, the hard-worked cart horse, and the lazy, over-fed family carriage pet, and we descend the scale, finding ourselves where we started, once more, at the *nutritive*, the purveyor of all the others; the first to develop in infancy, and the last to give up at the final dissolution. It may sound strange to some when I say that a person with poor digestion is generally "*superior*" to one whose digestion is remarkably good, and yet such is the case. It would not be so, however, were *other things equal*. Where digestion is weak, the brain is generally found to be strong and active, absorbing the forces of the nutritive system (and thus impairing the digestion), but a healthy digestion aids the activity of the brain and nervous system. A striking instance of this was the late Dr. Darwin, whose whole life was affected by the weakness of his nutritive system, which, however, did not hinder him from the most persistent and continuous mental labour and activity.

From what has been already said, the reader is prepared to hear that every member, feature, expression, or action, of an individual indicates something either for or against "superiority." Let us then examine some of these, as in every one we shall find valuable hints to aid us in determining "superiority."

The head is an epitome of the individual—we will first consider its size. The circumference of the head is anything but an accurate criterion of "superiority." Size of brain, and also shape,—towering or flat, long or round, etc.,—are temperamental signs. A large cranium ever characterizes the lymphatic subject—large, smooth, and globular. Undoubtedly, *other things being equal, size* is a measure of power, but other conditions are seldom or never equal.

In judging of "superiority" by the head, nothing is more important than to observe if the brain increases or diminishes in size from the base to the apex. To illustrate this difference of structure, let a cocoa-nut be sawn in two about mid-way and the two pieces placed side by side, with the section downwards, and the difference in craniums will be well set forth. Although at the base they are precisely of the same size, the tapering one has less than half the capacity of the other. The contrast between the cranial development in A 7 and B 7 is very striking, and still more noticeable between A 2 and A 19. Development of the base, instead of the upper or superior portion of the brain, usually bespeaks grossness of organization, and hence "inferiority."

The Forehead.—A full, square forehead is favourable to "superiority;" and if the surface is uneven, with knotty protuberances, and bulging out over the eyes, it is of the highest type. Such a forehead betokens *life*,—a strong energetic organization; just as surely as a round, smooth forehead evinces passiveness and sluggishness of disposition. A 11 is a fair illustration of the former, and B 8, A 12, and A 14 illustrate the latter. Height of forehead is more favourable than width: A 4 has the former, B 8 the latter, and A 3 has both.

The Veins.—Where the veins, in any part of the body, stand out in bold relief, they show that there is activity in that locality; therefore, veinal prominence in the intellectual region is a good indication of "superiority."

The Brow.—A projecting brow is most frequently found accompanying a closely knit, energetic organization, and must be classed as tending to "superiority." An eyebrow amply supplied with hair

is also an excellent token. Hair upon the body
generally typifies strength, and is an element of
" superiority ; " but nowhere more so than on a
massive brow. Of this kind of brow, indicating
great " superiority," no finer instance could be
mentioned than the late Charles Darwin. Finely-
arched eyebrows commencing at the nose are not
always indicative of genius, but it would be extremely difficult to find a genius who had not such
a brow. A 1, A 3, B 15, and B 16 illustrate the
brows of genius. A 9 combines the projecting and
hairy brow.

The Eyes are not without their significance in judging of superiority. Together with their " setting,"
they perhaps reveal to an experienced observer more
than any other single feature, and may be taken
as almost pre-eminent, in determining real worth.
Activity, for instance, is one of the most essential
points in the estimation of " superiority," and this
quality the eyes reveal most unmistakably. A large
prominent eye must count unfavourably, in a vast
majority of instances; it indicates conversational
powers, which so often accompany a lymphatic temperament. Great talkers are proverbially little doers.
Even so long ago as the days of Aristotle, large and
protruding eyes were in disrepute so far as intelligence is concerned. Large, light, rolling eyes, I am
forced to believe, are undesirable ; and it is a curious
but a well-authenticated fact, that an exceptionally
large proportion of children possessed of such eyes
do not reach maturity.

Those having light, pale, or blue eyes are seldom

equal in force of character to those who have dark brown, black, or grey eyes; the latter usually found with red hair. A piercing, bright black eye is peerless in the strife for the supremacy. A small or sunken eye seems to bear no special relation to " superiority;" it is as often the accompaniment of inferior qualities as of superior. A medium-sized eye would appear, on the whole, to be most desirable. Distance between the eyes may generally be considered highly favourable to " superiority."

A *drooping eyelid* is an invariable sign of " inferiority," and frequently betrays it when, owing to the excellence of the general type, it might otherwise pass unobserved. The eyes of B 6, B 10, and A 15, impress me unfavourably, chiefly owing to their size; while those of B 2, B 8, B 11, A 7, A 18, and A 19, all add greatly to the sleepy expression of countenance, and detract equally from " superiority." Dr. Powell asserts that these drooping eyelids reveal the lurking presence of *lymph*, even in a slender frame; a statement, however, which I have not yet been able to verify. This defect is hereditary, but may be to some extent overcome.

A long, silky eyelash, though very pretty and poetic, seems to indicate the presence of disease, more or less active, and generally of a scrofulous or consumptive tendency: it must be therefore classed as unfavourable to " superiority."

The *nose* is a most important feature in estimating " superiority," as well as in determining character generally; as may be inferred from the fact that a false nose generally makes a most effectual disguise. The face may be divided into three sections—upper, middle, and lower; or forehead, nose, and chin. The nose should constitute a full third of the length of the face; in fact, it may almost be said, the longer it is the better, for a long, prominent nose, if not fat, is proverbially a mark of power, or "superiority." Glance at the portraits of those whose fame and genius have become historic, and note how well-nigh impossible it is to find one having a short nose in a collection of hundreds. "That fellow has nose," the Spaniards say, when any one displays marked ability. The expression may be uttered with little forethought, but the very existence of such a proverb tends to show that observers in every age have been impressed with the value of the nose in indicating character. Napoleon said: "Give me a man with a large allowance of nose, for in my observations of men I have almost invariably found that a long nose and a long head go together."

When the nose is viewed in profile, the point should project to about one-third of its length: the dorsum or ridge should be broad, with well-defined sides throughout the whole length. A Roman or aquiline fulness of the dorsum, with but slight, if any, indentation at the point of union with the forehead, an increasing broadness between the eyes, and uniting with the orbitary arch in a true and graceful curve, are all unmistakable traces of "superiority." Lavater, speaking of such a nose, says: "It is worth more than a kingdom." Small nostrils he considers "a sign of unenterprising timidity."

The nose, like the face, may be divided into three sections, the upper third corresponding to the mental, the middle to the motive, and the lower to the animal. An excess in the *lower* portion, as in A 15 and A 19, generally shows that the animal or lymphatic element predominates. When the *middle* section preponderates, it denotes executive ability, which must be reckoned as favourable to " superiority." We admit that this form of nose is frequently met with in persons having small heads, and who are almost destitute of intellectual and moral taste ; the principle, however, holds good as far as it goes : they have great energy of character, but as they possess none of the higher faculties, they must naturally fall far behind when we come to sum up their superior points. The *upper* nasal section largely developed, as in A 1, A 2, A 3, and A 9, should be counted as an indication of mental strength. It is a mark of superior power, and frequently of genius, but I fear it is not so reliable a criterion of the mental powers as are the other portions indicative of the motive and animal.

A prominent nose is of itself an excellent feature, and generally denotes a person ready for great enterprises and achievements, just as surely as a flat nose betokens an ease-loving, lymphatic nature : the North American Indian is a tolerably good illustration of the former, while the African displays character as shown by the latter.

The *mouth* and *lips* are very expressive, but they are so flexible as to pertain largely to pathognomy. Lavater, after teaching nearly a score of pupils for three years, gave up in despair, not finding one who was capable of perceiving the half of what to him seemed so apparent in these features. By these alone, however, we cannot determine "superiority." Certainly, firm lips are coupled with a corresponding character—presumptive evidence of "superiority"—and yet they may belong to a small-souled individual, whose sole distinguishing trait is obstinacy (A 20). Again, weak lips with quickness of motion are said to declare a weak and wavering character, and yet, from my standpoint an individual thus endowed may be incomparably "superior" to a person with firm lips. One may have a noble, generous, whole-souled nature, and yet lack *decision of character*, which is more especially manifested in the lips. Immanuel Kant, one of Germany's greatest philosophers (A 7), does not display a tithe of the character about the mouth which is to be seen in A 5 and A 10.

A very full and fleshy under lip usually indicates sensuality and indolence. Of the two extremes—a large mouth and lips, and small ones—the "superior" individual will *four* times out of *five* be found with the latter. Calm lips, well delineated, well closed, indicate consideration, discretion, and firmness, and constitute the happy mean, the highest type, the true ideal. The mouths of B 3, B 7, A 19, and A 20 may almost be ranked as deformities. A 1, A 2, B 4, B 12, B 15, and B 18, all have mouths that may be considered perfect.

The Teeth are to some extent indicative of character. Large, long, projecting incisors almost invariably point to "inferiority." Small teeth are not unfavourable to "superiority:" those of a medium size are probably, on the whole, the most desirable.

The Chin.—A full chin has long been considered to betoken something positive, and a retreating chin tells of negative qualities; but this cannot be accepted without considerable qualification. A heavy under jaw would not alone make a Napoleon, yet all men of his character are sure to have massive chins. A large chin is favourable when accompanied by a large nose and forehead; but this combination is so seldom seen that it is safer to say that a moderately full chin is the most desirable of all. Character is apparent in the chins of A 1, A 8, A 9, and A 10, although the chin of A 9 is small in proportion to the other features.

The Face, Cheeks, and Neck.—A few words on the cheeks and neck, together with some general observations on the features collectively, will complete this part of my subject. The primary function of the mouth, jaws, teeth, and cheeks being the mastication of food for the support of the body, any excessive development of these features indicates a preponderance of the nutritive system, and consequent " inferiority," as explained at the beginning of this chapter. A neck large in comparison with the cerebral development above it—especially when the moral and intellectual lobes are weak—denotes brutal " inferiority." The necks of A 8 and B 7 are as fully developed as is desirable, while those of A 11, A 14, A 15, A 16, A 17, and A 18 show excessive development.

The face of a well-balanced individual should be of a somewhat oval shape, as B 15, B 16, and B 18. Such may be called the original form and proportions of the "human face divine." Nevertheless variations of or *deviations from the true model* appear to work little harm, but, on the contrary, a positive benefit; they seem indeed to be an almost indispensable accompaniment to the achievement of eminence, or even success in life. But such variations have their limits, which cannot be passed without serious detriment; and these general considerations should ever be borne in mind when estimating "superiority."

For instance, we will suppose a well-balanced individual to devote himself almost exclusively to mental pursuits, and to neglect his physical culture. He may attain eminence, and experience no serious inconvenience; but to his children he transmits organs strong or feeble in proportion to their activity in himself. Such children appear with towering brain, while the features of the lower portion of the face are more or less weak and insignificant; thus exhibiting a deviation from the true type which is positively detrimental. Parents with as high an intellectual development as seen in A 2, A 6, and A 7, seldom transmit to their children a desirable endowment either mental or physical; the only factors which may materially modify this result are the condition and tendencies of the mother.

The strongest human combination that we can conceive is that in which the three sections of the head—the animal, executive, and mental—are all developed to their full capacity, making the face square, or of the Napoleonic type.

Of the remaining portions of the human frame little need be said. A short-fingered, plump, soft, baby-like hand is in a large majority of instances unfavourable to " superiority," as is also the thick-set body it is almost sure to accompany. Still more unfavourable is it if the hand usually feels cold, clammy, and moist to the touch of another, as this condition unmistakably indicates the presence of excessive lymph. Long, slender hands or feet accompanying a similar build may be counted as favourable.

A hollow forehead or nose, an open mouth, long, thin, and tender, fair hair, and a shrill, timid voice, all indicate " inferiority." All roundness of features or muscles is disadvantageous, as it savours too much of the phlegmatic and lymphatic temperament. Angularity should be counted on the side of " superiority," as also should fineness of quality.

The strong contrast in the texture and quality of the first half-dozen couples in this work, as well as in the last four or five, is most noticeable. A 14 is of a fine, sleek exterior, yet the quality is defective, as is that of A 15. Such fine organizations are worth little without " activity."

Activity is the principal consideration in determining " superiority," and in assisting the reader to judge I would emphasize the necessity of noticing two or three of its phases, such as the motives or conditions which are likely to awaken it, the proportion of activity to inactivity, and whether it be physical or mental, or both.

Physical activity, being always visible, is easily estimated, but mental activity is frequently occult, and hence one is very liable to err in estimating it. Many persons are *spasmodically* active, like beasts of prey, while others are *incessantly* active, and, therefore, correspondingly " superior." Physical activity is vastly inferior to mental; activity in grappling with moral and philosophical themes is of the most favourable stamp. Professor Alexander Bain, in an essay on the " Correlation of Nervous and Mental Forces," labours to analyze activity. He says :—

" Great activity and great sensibility are extreme phases; each using a large amount of power, and therefore scarcely to be coupled in the same system. The active, energetic man, loving activity for *its own sake*, moving in every direction, wants the delicate circumspection of another who does not love activity for its own sake, but is energetic only at the spur of his special ends."

I cannot accept the idea presented in the first sentence of this quotation, but I heartily subscribe to that of the second, provided it be conceded that the " special ends " are sure to be so numerous as to " spur " the individual on unceasingly. The latter has a predominant cerebro-nervous system, and is almost incomparably the " superior "; the former has the larger organic system, larger ganglia.

Activity of speech is usually favourable, but not always so. Full eyes—indicative of conversational powers—I have already mentioned as generally telling against "superiority," and yet persons with such eyes are sure to be great, if not rapid, talkers: it is not so much the utterance of *words* as of *ideas* that counts,—the expression of thoughts, not the rehearsal of idle gossip. Activity of step is generally favourable.

Activity is then, I repeat, a most important factor in the determination of "superiority," for, being evidence of high *quality*, it frequently makes up for considerable deficiency in *quantity* of brain. Again, there are many persons who, in spite of frames physically far from robust, and a general lack of that indefinite physiological force we call *vitality*, must be considered as having very decided elements of "superiority." In them the intensity and activity of a relatively small brain more than make up for the lesser quantity, but the intense, almost morbid activity of the nervous system draws to itself all the nourishment of the system, and impoverishes other parts of the body, often giving rise to weak health. The poet Shelley may be taken as a type of this organisation, in which the active brain makes too heavy a draft upon the vital forces of the body. Yet such persons, unless actually in ill-health, will be most frequently the "superior" in the sexual scale.

As to the arterial circulation being an indication of activity and "superiority" generally, I have not been able to satisfy myself, but it is one of the conditions which is worth the attention of investigators.

I have sought in vain for some sure criterion of activity which might enable one to speak with certainty on the point of "superiority" upon seeing a portrait. Slender muscles, a sharp nose or piercing eyes, generally indicate activity; but these are not infallible signs, and to judge accurately one must see the living subject.

The illustrations at the end of this work comprise twenty pairs, and the general design has been to arrange the males in a descending and the females in an ascending order; hence, the first pair represent the highest type of man and the lowest of woman, while the last pair represent the very reverse. The first five couples would produce an enormous excess of girls; the last five, a similar excess of boys; and the middle ten, taken altogether, about an average o! the sexes.

We will now turn our attention to the comparative merits of the respective couples. It must, however, be borne in mind that in comparing these individuals I am only trying to gauge approximately that inherent, subtle, natural force which as we proceed will be found to vary so considerably.

No. A 1 is a model figure of manhood. It is hardly necessary to discuss the different features of that noble face, for they are one and all perfect.

There are doubtless lower types of womanhood than B 1, but they are difficult to find; everything about her is gross and repulsive; nearly all her features are in striking contrast to those of the man by her side. Assuming it to be possible that such a man could ever marry such a woman, what would be the sex of the issue?

As a matter of convenience, let us represent the average human being by 100. According to this standard, A 1 should rank at about 130, being fully 30 above an average specimen of his race. B 1 should not be rated at more than 50; hence the disparity between them. Were they to have *twenty* children, every one would most surely be a *girl*, for by no possibility can 50 outweigh 130 in Nature's impartial scales.

No. 2 shows us another couple between whom there is likewise a great inequality. I should rate the man at about 130, and the woman at 50. She is lower than B 1, in many respects; has far less activity or life. A has the advantage of her in every particular except as regards age, which perhaps reduces the difference between them by about 10. Her drooping eyelids show her "inferiority." She may have half the quantity of brain which A has, but his is worth fully ten times as much as hers. Daughters would invariably result from their union.

No. 3 gives us another matchless wit who should rank as high as either of the foregoing. The woman at his left we will estimate as high as 80. Her youth and spirits reduce the difference between them still more, say by 15 or 20. The upper portion of her head is good, but her head as a whole is more ornamental than useful. She was born for the kitchen, has always lived there, and has no ambition beyond: but who can gauge the capabilities of that ever active brain of his? A long debilitating sickness might reduce him below her standard; so that a son—a delicate one, be it remembered—might be born to them, but even that is not likely to occur.

No. 4 represents a very able, active man and a young woman, or squaw, of no very mean natural endowment. We will rate him at 130 and her at 90; her youth gives her a further advantage of perhaps 15. He apparently abounds in everything pertaining to life, thought, energy, and zeal; and she is capable of almost anything but high mental and moral efforts. How much superior her features are to those of B 3! She may have less brain, but it is certainly of a far better quality. A and B would most certainly have daughters, yet the disparity between them is not so great as in the preceding cases.

No. 5 presents a man with a keen, cultured, intense organization, past middle life, and a young woman who is unlettered, commonplace, and simple. I would place him at 130, minus 20 on account of his years; she can hardly be rated higher than 80; hence, a family of daughters would inevitably result from their union.

No. 6 presents a man whose somewhat dull expression of countenance betokens an excess of brain. I would estimate him at about 125. The woman certainly has a most voluptuous eye, and an ease-loving disposition. I cannot place her at more than 85. I rate her lower than the squaw, even though she be a person of noble birth and royal blood, for the stern necessities of that child of the forest have developed her nature, while the easy, luxurious life of the royal lady has steadily enervated a frame predisposed to sensuality. Daughters would surely be the result of this union, owing to his vast "superiority."

No. 7 gives us a profound philosopher mated with a sturdy dairywoman. He has nearly twice her volume of brain, yet hers is the more active, or rather, her entire organization is more active than his, and if the expression here given us be true to life, his mental processes must be very slow. As they appear before us, I would class him at about 110, and her 100; daughters would under ordinary circumstances be the almost certain fruit of their union.

No. 8 shows us a man charged with zeal, activity, and nervous force, and although the quality of his organization is not as fine as some of the preceding ones, his energy surpasses that of most of them, and he should at least be rated at 110. The woman is of the nervous lymphatic order, and the perfection of the musical temperament. She will become as nimble and vivacious as you please for a few minutes when called to play upon her favourite instrument, and then sink back into about the attitude and mood in which she is portrayed. She is, on the whole, inferior; I estimate her at 95. She has the advantage over her partner in regard to age, but we must nevertheless count upon daughters only.

No. 9 sets before us a remarkable type of manhood —a man indeed who is organized to undertake any great and daring enterprise, and who will carry through whatever he undertakes. His brain is wonderfully compact and intense, and were there more of it, he might become a terror to his race. He should be estimated at about 120. Behind him stands a good matronly, managing woman: her brain seems to be about the size of his, and quite as fine, but by no means so intense; yet she will pass for a fair average of womanhood, and we accordingly class her at 100. They are more nearly peers from Nature's point of view than from the world's. Few, if any, sons would be likely to gladden their hearth.

No. 10 exhibits a man with a good, solid, deliberate, constructiv head, mated with a commonplace, plain-looking girl. Owing to his want of activity we can hardly place him higher than 115, and from that we deduct a discount of 20 on account of his years, he being at least as old again as she is. She should be rated at about 100, and consequently more sons would be born to them than daughters, owing to Nature's finding in her more of the conditions of "superiority" than in him. His vital forces have grown sluggish, while hers are at the flood-tide of youth. His mind is well stored with the accumulations of mental toil, yet I cannot imagine a very active brain accompanying a substantial nutritive system : those massive cheeks and that double chin show that the stomach is inclined to prepare more than the body can find use for.

No. A 11 represents a man with a powerful organization : his brain is large, though the quality is rather inferior, or coarse, but the zeal or activity of his nature atones largely for this defect : we will estimate him at 115. The woman by his side is possessed of talents which render her popular, but her drooping eyelids reveal the fact that she is of a sluggish nature, and she cannot therefore be classed above par—100. The difference in age is slight ; they would have daughters.

No. A 12 is a man who manifests solidity throughout ; accumulation is a ruling principle in his nature, quite in contrast with A 9. He has a large brain of fair quality, but his activity is deficient; he might be placed as high as 110, but the discount on account of age reduces him below 100. The lady at his left is of much the same general temperament, has about the same quality of organization, a less quantity of brain, but her activity far more than compensates for this, and renders her decidedly the "superior" —say about 115. Their union would result in sons.

No. A 13 presents stolidity no less than solidity, yet there is no trace of brilliancy. The quantity of brain is ample, quality barely average, and activity quite ordinary; we will rate him at 105. The lady seems to have a good cerebral development, perhaps three-fourths the quantity he possesses; the quality of hers, however, far surpasses that of his, and her activity is much greater than his. She should be placed at 120; and as his age is against him, he must rank below 100; therefore they would have a family of sons.

No. A 14 is one of those abnormal human structures which, in some respects, it is difficult to describe. Shall we call the organization coarse, when the flesh is as white and soft as a baby's? He would be pronounced as of a lymphatic tendency, yet I cannot explain why, with his enormous bulk, he has a bright eye, so unlike B 2, when both must be classed as lymphatic. Yet I know this, that he, notwithstanding his bulk and age, is " superior " to B 2, with her youth and drooping eyelids. B 14 is a fine type of womanhood; quantity, quality, activity, and age are all about as favourable as could be desired. I should class him at 75 or thereabouts, and her at 120. Sons would invariably be born to them.

No. 15 A exhibits nothing at all felicitous. I do not think that the faces of these two individuals could retain their contrast for a single month, were their habits to become in every way identical. She has probably more brain than he, is of a much finer organization, and is far more active. I think Nature would rate him at about 75, and her at 125—the result being sons.

No. 16 A presents a plump, pleasant face, which is almost sure to be florid. I have seen many persons with faces of that type who were very quick in all their bodily movements, and in their speech and thoughts also; yet I have never met with one who was a profound thinker. The activity of such men is physical rather than mental; their thoughts are superficial. His eye betrays mental acumen nevertheless, and we may be justified in placing him at 90. B 16 is almost an ideal type, and we readily estimate her at 125. Boys would surely result from such an alliance.

No. 17 A presents another case of abnormal development of the nutritive at the expense of the cerebral system. If the brain of A 5 were put into his head, his superabundant flesh would be rapidly converted into nerve food. His partner is a noble specimen of womanhood; but, owing to a moderate digestive power, her brain consumes every particle of available vitality in her body, and still calls for more. Estimated by Nature's standard, she would rate as high as 130, while he could not exceed 75. Boys would, of course, be the result of their union.

No. 18 A has a weak head, a weak stomach, and

is weak in every respect. He should rank below 50; while that young, keen, closely-knit, highly-organized being at his side should be placed at 130. Sons would be *inevitable* in the event of their union.

No. 19 A presents a specimen of the lowest type of humanity, hardly rating at 45. The ideal type at his left should surely be classed at 125.

No. 20 represents if possible a still further extreme, and after what has been previously said will surely need no comment.

The "superiority" we have been considering in this and the preceding chapter implies a difference between the parents, and if it has, as I claim, an effect upon reproduction, this effect ought to be more pronounced if the difference is increased. Let us then turn to a more general aspect of the question, and note what effect is produced by differences greater than those constituting this bare "superiority."

In speaking in Chapter V. of sterility, reference was made to the facts of human hybridity as recorded by M. Broca, in his work on that subject. It is shown by him that sterility in the crossing of races seems to be affected by the amount of difference in their organization—the white race with the Australians being generally infertile, and the white race with the negro, in the United States, showing a tendency to unilateral hybridity. This tendency is also shown by Waitz to exist in the union of Arab men (of the Semitic or white race) with the negresses of Darfour, whose children, he adds, are feeble, and soon die out. Dr. Nott gives the result of his experience in saying that the mulattoes are the shortest lived of any race, and M. Broca mentions the circumstance of the rapid dying out of the Lipplappen, the mulatto race of Java sprung from Dutch fathers and Malayan mothers. The same authority says that the mulatto women are peculiarly delicate and subject to disease.

In further illustration of this, Barrow says, in his "Travels in South Africa," published in 1801, that the Hottentots, then a numerous people, impressed him, by their languid habits and dejected temperament, with the idea of an effete race whose vitality was all but exhausted. This he attributes to their being a hybrid race, resulting from an intermixture of former Chinese and Malayan immigrants with the black aborigines. The form of the skull and other characters led him to connect them with the Chinese, whose characteristics were familiar to him, from his having been a member of Lord Macartney's embassy to that country. On the strength of all this he asserted his belief that they were destined to speedy decline and extinction. The event has proved that he was right, for to-day the Hottentots have died out, and the hybrids of these with the white and other races are rapidly decreasing.

Now all this clearly supports my theory, and *race* might be given as one factor in the determination of sex; but it seems needless, as we are principally concerned with sex in single nations, and, moreover, difference of *race* is only an extreme form of difference of organization.

From these statements we seem to learn that nature prefers the union of parents approximately equal to, though, it may be, very different from, each other, and to such she gives most offspring, with the sexes equally divided; and a normal household of this kind undoubtedly has most of the elements of domestic happiness. But in proportion as the difference between the parents, whether temperamental or racial (that is, constitutional and functional), becomes greater, so do their children become more and more abnormal, being reduced in number, and possibly in vitality, and being more and more exclusively of one sex, namely, that of the inferior parent. When the inequality is extreme, we have comparative sterility, or, indeed, in many cases complete sterility, either in the first cross, or in the second or third generation. That it is this inequality alone which produces the infertility is shown by the fact that the women of low race, who are infertile with the whites, are fertile with men of their own race ; and the same may be said of women of high race in union with men of lower human types.

Now that which is true of races is also true of individuals, and this possibly explains many cases of sterility in our midst, especially in the families of eminent men, who very often have few children or none at all, and whose posterity frequently die out in a generation or two. And to account for this sterility we need not always assume very great " inferiority" on one side. Considerable divergence may co-exist with excellence in different ways, but in most cases one parent will doubtless manifest a decided physiological " superiority." It seems, however, that great " superiority " on the side of the father is not so effectual a bar to fertility as an equal " superiority " on the mother's part ; at least the above-mentioned phenomena of unilateral hybridity appear to prove, or at any rate suggest, this conclusion.

To state the case concisely, assimilation is most perfect between individuals physiologically equal, and becomes more difficult at every successive stage of inequality between the parents till the point of absolute sterility is reached.   It may be noticed that this result has a marked bearing on the origin of species by Natural Selection, as Darwin evidently considered (" Origin of Species," ix.), by tending to the perpetuation of pure breeds when once a variety has succeeded in establishing any considerable difference from the parent stock,—the mongrel offspring being produced somewhat less readily than the true-bred issue of the new race ; and it will also explain why mulattoes are short lived and subject to disease, just as hybrids between varieties of plants when the varieties differ at all widely.

A remarkable corroboration of the above is furnished by Dr. Nott, who found that the offspring of negresses and whites of the dark or southern type were more prolific and more vigorous than those born of negresses and Anglo-Saxon or northern whites.   Here there could be no question of the " superiority" of the northern over the southern white. They are probably perfectly equal, but the southern or Celtic white is somewhat less divergent from the negro race, constitutionally, than is the fair-haired northern or Teutonic white ; and on this account, though the " superiority " of the white father is as great in the one case as in the other, the lesser divergence, constitutionally, in the case of the southern or Celtic type renders the crossing more fertile.

And here with regard to the difference in constitution of species being the cause of their sterility when crossed, and as to the individual difference producing differences of fertility among individuals of the same species, it may be well to cite the following from Darwin in the " Origin of Species " to show that I am advancing no scientific heresy, but a doctrine fully admitted as a general principle, though not as yet applied to the human race :—

" The hybrids raised from two species which are very difficult to cross, and which rarely produce any offspring, are generally very sterile, but the parallelism between the difficulty of making a first cross and the sterility of the hybrids thus produced . . . is by no means strict. . . . The fertility of first crosses is innately variable; for it is not always the same in degree when the same two species are crossed under the same circumstances,—*it depends in part upon the constitution of the individuals chosen.* So it is with hybrids, for their degree of fertility is often found to differ greatly in the several individuals raised from seed out of the same capsule and exposed to the same conditions.

" By the term ' systematic affinity' is meant the general resemblance between species in structure and constitution. Now the fertility of first crosses and of the hybrids produced from them is largely governed by their systematic affinity. This is clearly shown by hybrids never having been raised between species ranked by systematists in distinct families, and on the other hand by very closely allied species generally uniting with facility. But the correspondence between systematic affinity and the facility of crossing is by no means strict. . . . No one has been able to point out what kind or what amount of difference in any recognizable character is sufficient to prevent two species crossing."

Now this latter is precisely my point,—that difference in organization affects the reproduction of species, and, in a lesser degree, of varieties and of races; and further than this, as Darwin says, even individual differences produce effects *precisely similar in kind*. I only add that the first manifestation of this effect is the casting of sex; that is, variation in the constitution of individuals will produce variations, first in the sex, and afterwards in the vigour and number of offspring, and if the difference be increased will finally result in sterility.

Darwin makes his point still clearer a little further on by saying, " The foregoing facts appear to me clearly to indicate that the *sterility both of first crosses and of hybrids is simply incidental or dependent on unknown differences in their reproductive systems.*" And again, " Gärtner found there was sometimes an innate difference in different *individuals* of the same two species in crossing."

Thus it is clearly recognized that this result of difference in organization exists as a general rule. Darwin is speaking of lower animals and plants, but in the above cited cases of *human* hybridity I have endeavoured to give it a particular and more minute application, in order to show that the theory of sex which I advance is founded upon a well-recognized principle in science.

Hitherto I have chiefly spoken of hybridism (*i.e.*, a crossing of *species*), and some may object that the case is quite different with mongrels (*i.e.*, the result o   crossing of *races* or *varieties* of the same species);

but if we accept the origin of species by natural selection, there is no fundamental or essential difference between these,—the difference between individuals, races, and species being one of degree only. And Darwin adduces abundant evidence to show that there are great differences in the fertility of races when crossed, which is equally cogent whether we accept the doctrine of evolution or not.

The main result is the same,—that divergence of organization produces difference in fertility, such fertility decreasing as the amount of divergence increases until absolute sterility is reached.

We have now arrived at a point from which we may survey the progress made. In the light of this and the three previous chapters, I may place my theory before the reader in a few words, by way of recapitulation and elucidation, and I will in the following chapter proceed to establish its truth by application to nature, and by verification in fact.

I assume, then, that the sexes are ideally equal, but that in certain cases, owing to predominance or defect of some part or section of the organization, one individual is very frequently the permanent " superior" of the other, and even where husband and wife are nearly on an equality, some slight causes will usually operate to produce in one or other a temporary "superiority," which will give to that parent the casting of the sex of the offspring. Owing to a law, recondite and as yet inexplicable, but analogous to the law of polarity in magnetism and electricity, the " superior " parent determines the sex of the offspring, which will be of the sex opposite to that of the " superior " parent.

Among the determining causes of "superiority," cerebral development and activity hold the first place, and in order to ascertain the nervous status of an individual, regard must be had to the manifestations of energy considered in this chapter, of which temperament is perhaps the most important. To these must be added various transient influences, the most important of them being the state of health. Physical vigour, habits of life, state of nutrition, mental condition, circumstances, recent occupation— all are factors in determining the "superiority" of an individual at different times, and of course the mere lapse of time, and the consequent change of constitution, will have an important influence. Seniority in age is usually an element of "inferiority"—and, other things being equal, the younger parent will assuredly be the "superior," and will therefore produce the opposite sex in the offspring.

CHAPTER VIII.

VERIFICATION FROM OBSERVED FACTS.

THERE is no better way to elucidate a principle than to relate one's experience in arriving at it. I will therefore briefly point out the difficulties I had to encounter in my investigation of this question, and how I was guided to my final conclusions.

In my own household I first became personally interested in the question of sex. I sought information from various sources, and undertook to examine, in turn, many of the most plausible of the current theories that presented themselves; but on investigation, for the most part I discovered reasonable grounds for their rejection. After a time, I found that I had almost unconsciously fallen into a habit

of meditating upon the law of sex, and in order to
furnish some data on which to base a conclusion, I
observed the number of either sex in the various
families of my acquaintance and compared and
studied the relative characteristics of the parents
individually. At last, amid the chaos of facts,
I began to discern a slight semblance of order,
though hardly sufficient to warrant the construction
of even a preliminary hypothesis; this had the effect,
however, of leading me to a settled determination to
solve the problem, and so intensified my zeal that
my mind fairly refused to dwell upon any other topic.
Feeling confident that I had adopted the true
method to ensure ultimate success, viz., that of
scrutinizing every fact presented, of observing before
theorizing, I at length devoted myself wholly to my
subject. I now searched through works on political
economy, sociology, books of genealogies, biographies, medical literature, life statistics, census reports, baptismal records, etc., in order to find any
facts that might throw light upon the desired point.
I consulted physicians, but, as a rule, they had
nothing to offer save one or more of the many
current theories. I very soon learned to test the
possible truth of any of these by simply ascertaining
whether they contained any *principle of balance*.
After rejecting a number of them for want of this
principle, I was led to adopt, as a basis, the assumption
that *each sex produces its opposite;* for I now per-
ceived that a theory reared on this basis could not
fail to contain the indispensable principle of self-
adjustment. I then set about observing and comparing facts. I sought out large families in which

the children were all of one sex. I made their
acquaintance, and studied the features and characteristics of the parents. I soon found instances favour-
ing the hypothesis which I had adopted as my basis.
Whilst pursuing my investigations, it so happened that
of two families of girls that came under my notice,
the fathers had Roman noses, and the mothers had
noses somewhat inferior in type; while in three
families of boys, these same nasal conditions of the
parents were reversed. I now made the nose the
chief object of study, and discovered that the more
decidedly Roman the father's nose, the larger was
the proportion of daughters in his family, and that
the converse was equally frequent. In cases where
the noses of parents were nearly of similar type,
the sexes in their families were more nearly balanced.
For many months I was half disposed to conclude
that I had discovered a definite and almost unmistakable guide. I next carefully prepared lists of all
the married persons I had ever known, with the
proportion of the sexes in their respective families.
This proved a most invaluable help to me, for whenever a new idea occurred I could readily test its
worth by glancing over my list and recalling the
tabulated couples.

As I extended the range of my observation, and
made new acquaintances, I met with couples the sexes
of whose children were distributed in a way wholly
at variance with my rule ; and this discovery left me
almost as much in the dark as at first. I felt that
with only a single clear exception to my nasal rule I
might still speculate, but never settle the question :
however serviceable the nose might be as an index, I

plainly saw that it was destined to prove an unsafe guide if taken by itself. I saw, in short, that the problem was not quite so simple as I had imagined.

I soon after noticed that the parents with these aquiline noses were the ruling spirits in their respective households. Moreover I found that in numerous families where neither might be said to rule, the number of boys and girls was pretty equal. On one occasion, hearing a lady, in the course of conversation, say that her father's family was composed of *six* daughters and *one* son, I ventured to tell her that her father governed the household. She was surprised, but replied at once, " Nothing could be more true," and desired to know the source from whence I derived my information,—little dreaming that it came from her own lips. Commenting upon her parents afterwards, she inadvertently admitted that her father was in every way superior to her mother, and qualified to fill a high position in life. This was all I could have desired. I had now found so many confirmatory facts of this nature that I was beginning to adopt this as a fixed theory. Several positive exceptions to this rule, however, presenting themselves, ere long, I was once more plunged into uncertainty. Those who have read the preceding chapters will readily understand how these general rules have their exceptions. But an aquiline nose and a governing disposition are indubitable elements of " superiority." Still there are other disturbing quantities, such as size, texture, and activity of brain, which may exist regardless of shape of nose or of the presence of any dominant trait: yet these two criteria are true in the *vast majority of cases, i.e.*, they are trustworthy indications of superiority of organization.

About this time, having already three valuable facts carefully observed, that a parent with a Roman nose, strong will, and ruling disposition *generally* has a majority of offspring of the opposite sex—I secured a fourth. My attention was arrested by the case of a family with whose genealogy for three generations I was quite familiar. A Roman-nosed father, of ruling disposition, had a large family, in the proportion of *three* daughters to *one* son. The sons died, but the daughters—intelligent and with aquiline noses—married, and their offspring averaged *three boys* to every *girl*. The present generation of this family has, again, three girls to every boy. I looked at other genealogies, and found, almost without exception, that where a family was all of one sex the succeeding generation showed a large preponderance of the opposite sex.

All this convinced me that I was right as to the main point, the existence of a principle of balance, and that the facts I had collected regarding the *nose*, *will*, and *authority* bore some intimate relation to the true law of sex, explained in Chapter XI. These preliminary successes encouraged me to prosecute my researches with renewed ardour.

The following cases which I found quoted in a French work on fecundity gave me inexpressible satisfaction and encouragement. They were taken . .m the observations of M. Girou, who during his lifelong search seems to have carried his explorations into the remotest regions. He says " they are facts for the curious to peruse, but destitute of any scientific value;" yet to me they had a deep meaning, and I only regretted the impossibility of seeing the living subjects. The fact that these cases seemed to M. Girou quite irreconcilable with any theory, and that they were indeed complete puzzles to him, gives them a high value as independent and impartial evidence. I take the liberty of italicizing the qualities which tell in my favour.

(1.) "M. A——, of a gay disposition, married a lady, *his senior*, gentle and melancholy, and more *corpulent* than he: *seven daughters* were the result." This is not a very clear case as it stands, yet gentleness of disposition and corpulency are not often found allied with Roman noses, strong wills, and ruling dispositions.

(2.) "M. B——, a man of *genius, large head* and slender body, with a *corpulent* wife of *moderate intelligence*, had *five daughters*." Viewed in the present clear light of "*superiority*," this case is all that could be desired: the *nervous* system as surely predominated in the husband, as did the *nutritive* in the wife.

(3.) "M. C——, of *extraordinary physical* and *intellectual activity*, and of rare *tenacity*, married a lady of a soft, *nonchalante* disposition. They had *eight* daughters and *one son*." This, too, is a perfect illustration of my theory, clearer now, however, than it was years ago, although "tenacity" at once indicated will, in the father; and this was one of the points I had then established.

(4.) "Mme. A——, a strong woman, with masculine voice and slightly bearded chin, had seven children, all sons." Here is energy of will beyond any doubt, and although no idea is given us of the husband, the case, on the whole, supports my view very satisfactorily.

(5.) " Mme. B—— a small, *dry, bilious* woman, had *twelve* sons and *two* daughters in the course of eighteen years." Here again we have very incomplete data, but such women seldom lack wills, or other marked traits of character.

(6.) " M. D——, of a sweet and *tranquil* disposition and *well-nourished* body, married a *very nervous, active,* and *intelligent* lady." If any of my readers have doubts as to the sex of the *thirteen* children born of this union, they have not read the earlier pages with attention. All were *sons*, of course, and a more fitting selection for securing such a result could hardly be made. How the parents might, with a knowledge of the law of sex, have also had daughters, will be explained in Chapter X.

(7.) " M. E——, a small man, *wiry, dark-complexioned, active,* and of strong passions, had a wife *still more active* and *resolute*, and of a *dry temperament.* Ten sons resembling the mother, and one daughter, the image of the father, were born to them." This is a perfect illustration of my theory, fully developed.

Satisfactory and interesting as these cases now appear to me, they were anything but comforting to M. Girou. They, like thousands of others that might be collected, present a contradictory, chaotic jumble when contemplated in the light of any theory save the one which this book propounds. M. Girou impartially relates the facts as he finds them, notwithstanding that they so often tell against the

Hippocratic theory, which, in his perplexity, he falls back upon—that of comparative physical vigour.

The same indefatigable writer diligently studied the history of France from the eleventh century to the nineteenth, with all its noted characters, and found that:

(1.) " Men of great character, whether virtuous or dissolute, have had a predominance of daughters.

(2.) " Men of weak character have had a similar predominance of sons.

(3.) " All those who have married women of *great character* and *strong* wills have had an excess of *sons*."

The longer I dwelt upon the phenomena of sex, and sought to harmonize them, the more clearly I saw that by regarding Roman nose, strong will, ruling disposition, etc., as merely varying manifestations of a dominant vital force,—a general adjustment might be made.

To test this "superiority" on a large scale, it occurred to me that the United States statistical tables of the sex of mulattoes born in the Southern States—being mostly the children of black mothers— might almost substantiate or ruin my whole hypothesis. I was not then in a place where a census report could be obtained. I knew that there was a general excess of about 5 *per cent.* of *male* births, taking the whole population together, but I felt sure that should my theory of "superiority" be true, a large excess of *female* mulatto births would be found recorded. So anxious was I to examine the records on this point, that I forthwith undertook a journey of several thousand miles in order to gain access to them, and felt repaid upon finding an excess of from 12 to 15 per cent. of *female* mulatto children.

A most striking fact of the same kind, already alluded to, is cited by M. Broca in *Human Hybridity.* "The Lipplappen" (this is the name of the mulattoes of Java) "do not breed beyond the third generation . . . At the third generation *girls only are born*, which are sterile." This is a most striking confirmation of my idea that the " superior " parent determines for the *opposite sex*. The excess of females in the first generation (as in the case of the mulattoes just referred to) would result from the " superiority" of the white fathers, and this "superiority " would pass, in the second generation, to the sons of these mulatto girls. These sons would predominate and would be sufficiently " superior " to produce girls chiefly, or exclusively in the third generation, though these it appears are not vigorous enough to perpetuate the race.

As the law of sex in all its various bearings gradually became clearer to me, it was a source of never-failing interest to watch its manifestation in the crowded lecture-hall, in the streets, on the railway, as well as in the birth lists and obituary notices of the daily prints ; everywhere, in fact, where families can be observed or studied.

Upon the death of a gentleman, I heard an intimate friend of his candidly observe, "He was the *laziest* man I ever knew." This remark quickly attracted my attention, and I enquired as to his family. " Three *sons*," was the reply.

Having acquired the key to what I have termed " superiority," I was greatly encouraged by my ability to elucidate all puzzling phenomena accompanying disparity in age between parents. But it must not be supposed that after reaching this stage no shadows crossed my path, for even until very recently, facts have presented themselves which seemed to militate against my theory—though a proper investigation has never failed to make them harmonize.

I became acquainted with a man of good address, with a well-shaped head, and a Roman nose. His wife was of consumptive tendency, of literary tastes, but with an infantile nose, and an expression of general inefficiency. At that time, I had so much confidence in my ability to judge of " superiority" from the shape of the nose, that I did not hesitate secretly to decide—as was my custom—what the sex of their children must be, provided they had any. I said, therefore, " mostly girls, because of the father's superiority; " imagine my surprise when I learned that they had three *boys*.

A case of this kind invariably led me to attribute defects either to my theory or to my own power of observation, and to suspend all further research until the difficulty had been cleared away. A depression settled on my mind every time this family recurred to my memory—which was very often—but ere long all was satisfactorily explained:—Besides being extremely illiterate, the husband was in the habit of returning home in the evening in a state of inebriation, even prior to his marriage. The fact seemed incredible, and yet seeing him so, myself, several times, I could no longer have any doubt of his habits. Nothing more surely degrades the nature, destroys manhood, and counteracts "superiority," than intemperance.

An important chapter in the history of a family may often be read in the sex of the offspring. I met with the case of a "superior"-looking man with a wife of rather "inferior" type, whose family consisted of a daughter and four sons. This surprised me, and I sought for the cause. I learned that shortly after the daughter's birth the father was attacked by a severe and lingering illness, and had never been able to regain his former constitutional vigour and activity; in a word, it made him the "inferior" of his wife.

I can point to almost a score of families having two or three daughters, born during the period of the family's prosperity, who were followed by as many or more sons, ushered into the world in less fortunate circumstances, frequently accompanied by discouragement, idleness, and dissipation, the mother meantime becoming the ruling spirit, and the active sustainer of the household. Any one may recall similar instances in the circle of his acquaintances. "Nothing is more potent to depress the mind and disturb the harmony of the nervous system than failure in business." Recent lists published by physicians show that an astonishing percentage of the sick are those who have succumbed to ill-fortune.

In his "Constitution of Man" Combe introduces a few facts bearing on heredity, which I here make use of merely to show the sex produced under certain conditions and habits of life. I have never found anything to contradict them, save in the case of two or three dissolute characters, of *superior talents*, who had some daughters in spite of their *periodical* excesses. Instances like the following from Combe naturally served to strengthen my convictions (the italics are my own) :—

"In a case which fell under my own observation, the father of a family became ill, had a partial recovery, but relapsed, declined in health, and died in two months. Seven months after his death *a son* was born of the full age, whose existence was referable to the period of the partial recovery." The author proceeds to explain the idiosyncrasies of the child—he being "all mother," etc.,—and demonstrates the transmissibility of qualities. Can any who hold to the sexual theory of "*physical vigour*" explain why a *dying man* should beget *a son?*

"A friend told me that in his youth he lived in a country in which the gentlemen were much addicted to hard drinking, that he also frequently took part in their revels. Several of *his sons, born at that time*, although subsequently educated in a very different moral atmosphere, turned out strongly addicted to inebriety; whereas the children born after he had removed to a large town and formed more correct habits were not victims of this propensity.

"Another individual, of *superior talents*, described to me the wild and mischievous revelry in which he indulged at the time of his marriage, and congratulated himself on his subsequent domesticatoni and moral improvement. *His eldest sons, born in his riotous days*, notwithstanding a strictly moral education, turned out a personification of the father's actual condition at the time of their birth : and his younger children were more moral in proportion as they were removed from the vicious period. The mother, in this case, possessed a favourable development of brain."

I vainly regret that Combe is not more explicit with regard to the sex of the other children, but the case is quite satisfactory as far as it goes.

The cases of several large families, wherein the children were mostly of one sex, perplexed me greatly, till I learned to notice the drooping eyelid, with its ever-accompanying deficiencies.   See Chap. VII.

Since overcoming the difficulties above enumerated, I have never met with any individual cases not in perfect accord with my theory, but two or three general facts at one stage caused me much perplexity.

As occupations indicate to a considerable extent the intelligence of those who follow them, I soon learned to test my theory of comparative " superiority " by observing the proportions of the sexes amongst the different classes. By collating statistics from various sources, it will be found that Nature's distribution of the sexes is to some extent influenced by the occupation of the parents. In most rural communities the sexes of children are nearly equal— males slightly predominating. Among labourers in towns, artisans, and tradespeople generally, a much larger excess of sons will be found : these proportions will vary in different countries, according to their social customs : men are more dissolute than women in almost every grade of society, and particularly among the classes named. The man who breaks

stones all day and spends his evenings at the
public-houses evidently plays an inferior part in life
compared with his unfortunate wife, with all the
responsibilities of a household and family upon her,
from early dawn till late at night. So, too, the
mechanic who passes the day in watching the ceaseless revolutions of the lathe; or the engine driver
who oils the bearings, looks at the gauge, puts on
more fuel, and then sits down, does not make one-
fourth the mental effort required of his wife in her
endeavours to keep the children neat and clean, and
the home respectable. Tradesmen also have a large
excess of sons, and it will be found that this excess
varies in proportion to the amount of mental effort
required by the trade in which the father is engaged.
Bankers, contractors, solicitors, speculators, etc., as
may be supposed, have fewer sons—often a preponderance of daughters; while wine-merchants, small retail
dealers, and tavern-keepers, have more sons than any
other class. Philosophers, lawyers, editors, poets,
and literary men and brain-workers generally, have
a large excess of daughters. Clergymen, orators,
physicians, and musicians have not the preponderance
of girls which I expected to find; indeed, if extensive
and reliable statistics were gathered of these four
selected classes, I feel sure that all but the clergymen would be found to have an excess of boys: this
would, I think, be specially the case with musicians.
I need hardly add that lady musicians have an excess
of daughters. These latter facts seemed at first to
tell against my theory; but when I became more
familiar with the faculties essential for these pro-
fessions, and the temperaments which characterise

their exponents, I discovered that they afforded the strongest possible confirmation of my views. My present test of " superiority " now places some classes and callings in a far lower category than that in which I should at first have ranged them.

Mr. F. C. G. Neison, F.R.S., publishes statistics showing the relation between occupation and longevity in England, for the use of life insurance companies. It is scarcely to my purpose here to extract largely from these tables, since the average lives of coopers and carpenters, for example, can have little bearing on the question of sex; but a few figures may be of service.

Clergymen . . .	die	10	per 1,000 per annum.
Domestics and gardeners .	„	7·9	„ „ „ „
Beer dealers . . .	„	20	„ „ „ „
Wine „ . . .	„	23	„ „ „ „
Spirit „ . .	„	23·9	„ „ „ „
Inn and hotel-keepers .	„	26	„ „ „ „

(The greatest mortality of any known class).

Thus we see how largely the death-rate is affected by the habits of those of different occupations, and hence the effect such habits are likely to have on comparative " superiority." It must be evident from the facts already adduced, that the habits which result in such excessive mortality must also in a similar proportion tend to lower the standard of vitality, thus conducing to " inferiority; " and by extensive observation I have proved this to be the case. I find that in the families of clergymen the sexes are about equally distributed, the parents being about equally intelligent, sober, and moral; and the same rule applies to the second class mentioned in the table. On the other hand, the families of hotel-keepers, as already mentioned, show a large excess of sons, the habits of men of that class being such as to render them far "inferior" to their wives. It will thus be seen that there is considerable connection between these tables of mortality and my theory of "superiority."

Professor Bain, in his well-known work, "Mind and Body," speaking of the effects of stimulants such as alcohol, tea, coffee, tobacco, opium, and the like, most truly says: "They draw upon our vitality even till it is much below par: they are the large consumers, not producers, of vitality; they expend our stock of power in NERVE ELECTRICITY in a higher degree and with a more dangerous license than the ordinary stimulants of the senses." It is therefore evident that those addicted to the excessive use of these stimulants must forfeit their "superiority." Proof of this may be found in the fact that drunkards with temperate wives have sons almost exclusively. There *may* be individual cases where—from some clearly traceable cause—daughters are exclusively born to such parents, but I have never met with a single instance.

I have taken the trouble to make out a list of some *two hundred* business men of my acquaintance in whom the love of money-making may be said to be in the ascendant: they are the progenitors of some *six hundred and fifty children*, of whom *four hundred* are sons,—a proportion of nearly *two* sons to *one* daughter. Among fifty business acquaintances who have shown their mercantile incapacity by repeated failures, I find *one hundred and sixty* children, of whom only *seventy* are sons, or *seven* boys to *nine* girls. Selecting from these fifty men those of the poorest business capacity, although capable men in other respects, I find that their families are nearly all daughters; and conversely, on consulting a list of all the shrewd, grasping traders I have known,—those eminently successful in business,—I find they are the parents of sons almost exclusively; and I think it will generally prove true that business men have more sons than daughters. The faculty of acquiring wealth is not by any means the noblest element of our nature.

Although I have, in this respect, found a satisfactory confirmation of my theory in most cases, yet, owing to the boundless variety of matrimonial alliances, arising from difference in age, temperaments, habits, intellect, activity, etc., it is impossible to make sweeping statements that will admit of no exceptions. To say that no great business man has an excess of daughters would be pressing our theory too far into generalities, while almost every individual case requires to be judged by itself—the balance of "superiority" between the special pair being the essential point. Indeed I have two or three cases in my mind of prominent merchants with daughters only, and yet they are in harmony with my theory: they are instances in which the husbands are "superior" men, and besides an excellent business faculty, have other high endowments, while their wives are only of average ability. I only make the *general* statement that business men have more sons than daughters; and to test the accuracy of the assertion we must be governed by general averages, not by isolated cases.

The foregoing facts caused me no particular trouble to ascertain or to explain, but I was perplexed not a little in finding so many sons in the families of musicians, medical men, and public speakers, as already hinted: this difficulty I will now endeavour to elucidate.

Strictly speaking, music must be classed among the embellishments of life; leisure is essential to its practice or enjoyment. There are exceptions to all rules, but, as a class, musicians certainly have the lymphatic element in their nature. This in no way militates against the fact that they often possess fine, susceptible organizations, capable of *spasmodic* vivacity and exaltation. The musical temperament is a combination of the lymphatic and nervous, as in B 8; which is far " inferior " to a blending of the bilious and nervous, as in B 17. The renowned pianist, Gottschalk, well illustrates this point, as his drooping eyelids testify. In many instances of professional musicians of my acquaintance, I find no exceptions to this rule. On the contrary, I have long observed that a fondness for the piano is seldom accompanied by a love of close mental application or of active household duties.

What I have here said of musicians must be understood to apply to those who excel in performance of music, while manifesting no superior energy nor force of character in other directions. With such persons the receptive or passive and emotional elements are in the ascendant, and this betokens a lymphatic temperament, which we have seen is "inferior" in our sense. But here again I must warn the reader that this "inferiority" is not to be taken as implying any general want of ability. Once for all, when I speak of people as "inferior" in this matter of sex, I am far from wishing it to be understood that they are inferior in intellect. Persons of lymphatic temperament may be warm-hearted, generous, of fine tastes and sensibilities, clear-headed and clever in many ways, but they do not possess predominant cerebral energy, and they will have children chiefly of their own sex because "inferior" in casting sex.

But the case will be different with musical *composers*. The great masters of music equally with statesmen or poets are undoubtedly men of high and active original genius. Their cerebral power must be very great, for musical composition is assuredly a profound science, almost rivalling mathematics as a test of mental capacity. Such men take a very superior rank, and I should expect them generally to have a large excess of daughters, but as many of them have also much of the emotional element in their nature, the excess will be somewhat less marked amongst them than amongst men of more purely intellectual pursuits.

As to the medical faculty, they seem to sustain no uniform relation to "superiority,"—which indicates that doctors are *made* such more often than *born* such; a person without natural qualifications for a musician can seldom gain a livelihood by music, whereas fortune has a large share in determining the success of a medical man. Still in the families

of the *leading* men of the profession daughters will
be found to predominate.

Of orators I had always held an exalted opinion,
which probably in some degree accounts for my
astonishment at finding that my theory did not
place them in a very flattering position. Evidently
I had made a false estimate somewhere; either as to
the nature of the law of sex or of the oratorical
faculty, and an impartial examination showed it to
be with the latter. Mere public speakers do not
possess the highest order of faculties or intellect.
They are, in the words of Dryden, likely to be
amongst those " who think too little and who talk
too much." In a majority of cases the base of the
brain of a mere orator will be found to predominate
more or less over the superior portions, as in A 8,
A 14, A 15, A 16, A 17, etc.; and the lower section of the face to be amply developed. Orators
have prominent eyes, which indicate a ready flow
of language, and if they have not invariably large
mouths, the exceptions are rare. There is something in their composition allied to the musical
temperament; they are almost uniformly of a warm,
sympathetic, genial, social nature; they are rarely,
if ever, profound thinkers, philosophy and oratory
being seldom combined in one individual.

Here again I must explain that the above remarks
apply to real talkers—men who are noted for oratory
and little else. The case is very different with men
of sterling ability in other directions, whose genius
and culture naturally give them the power of forcibly
expressing their thoughts in addition to their other
qualities. This is notably the case with statesmen,

who must of necessity cultivate oratory, and who being men of ability, will probably excel in this as they would in most things to which they devoted themselves.

The results of the cholera plague in Philadelphia and Paris in 1832 proved a serious stumbling-block to me. In a preceding chapter I have dwelt upon this visitation sufficiently to convince the reader that no theory yet propounded explains the phenomenon of sex in the least, and a few lines here will show how I satisfied myself.

In the first place, I protest against all attempts to base conclusions upon limited data, for out of nearly 4,000 births in Philadelphia that year, there was an excess of only twenty-five girls. I have taken the births during the period of the epidemic only, and omitted those in the wards in which the disease did not appear, and find an excess of 17 per cent. of daughters ; but this would be a basis of ridiculously small dimensions from which to deduce an immutable law of the universe. A calculation of percentages may mislead grievously if based on a limited number of facts.

I believe that from my standpoint there could have been found causes sufficiently numerous to account for the 17 *per cent.* excess in female births. But at all events, I will mention what seems to me to have been the most probable predisposing causes: Every change in the mode of living is found to have an influence on the proportion of the sexes, as Dr. Napheys and others admit, and as all observation confirms. Now what precise modifications in the habits of the people were in-

duced by the presence of the epidemic it is difficult
to affirm positively, but they were evidently such
as would conduce to masculine " superiority " rather
than feminine, if we may judge by results.

We may reasonably infer that the men became
more temperate and abstemious in their habits, and
that the women, with their natural timidity and
strong maternal affections, were more alarmed and
depressed than the men; all these facts taken into
account would help largely to explain the case
But I strongly suspect there were other causes
involved, which, however, I do not feel bound to
demonstrate until such time as physicians are prepared to tell us a little more of the nature of
cholera.

The variations in quantity and quality of atmospheric electricity exert considerable influence on
the sensations and health of nervous and susceptible
constitutions. It has been proved that there are
two daily tides of positive atmospheric electricity—
the high tides between 9—12 A.M., and 6—9 P.M.;
the low tides from 2—5 P.M., and 1—5 A.M., varying
slightly in different countries. It is known that the
sick usually die during these ebbing hours. Dr. A.
Wislizenus has shown this to be the case, as may
be seen in the Transactions of the St. Louis Academy
of Medicine, and in Ferguson's work on Electricity.
The annual variations of electricity are as marked
as the diurnal. The quantity of positive atmospheric electricity is greatest in winter and least in
summer; it gradually increases in autumn and as
gradually declines in spring, bearing a constant
relation to the temperature. All are aware of the

susceptibility of fine organizations to atmospheric changes, which affect both their physical and mental activity.   It is also known that females are generally more susceptible than males to the extremes of heat and cold; and this, with the above suggestions as to change of habits, and their special fears as mothers, will probably be sufficient to account for the "inferiority" of the mothers, for the time, and thus account for the above mentioned 17 per cent. excess of daughters.

If any fail to see the importance of all this, they will doubtless clearly perceive it after perusing the following table, which was the last and crowning difficulty which I encountered in the whole course of my investigation.   It is indeed fortunate that I did not meet with this obstacle during the earlier stages of my investigation, for, in my impartiality and readiness —I had almost said eagerness—to hear and consider everything presented against my theory, this might perhaps have caused me to abandon the research altogether.   But with such a mass of facts already in my hands, all pointing to one conclusion, I made bold to look into these figures and to carefully scrutinize them : they are selected from tables compiled in Great Britain by C. O. G. Napier, F.R.S.

PROPORTION OF MALE TO 100 FEMALE BIRTHS.

		Per cent.
390	Parents of equal age . . .	91·8
276	Fathers one year older than the mothers	101·3
312	„ two to three years older .	101·8
211	„ four to six years older .	101·8
100	„ six to ten years older .	120·1
168	„ ten to sixteen years older .	144·3

Per cent.

120 Fathers	seventeen to twenty-five years older than the mothers	189·7
80	„	twenty-six to thirty-two years older .	125·6
45	„	thirty-three to forty years older .	112·6
18	„	forty to fifty years older (mothers under twenty-five)	115·4
13	„	forty to fifty years older (mothers over twenty-five) . . .	91·6

MOTHERS OLDER THAN FATHERS.

88 From	one to three years older	. .	94·3
77	„	three to five years older	.	88·8
66	„	five to ten years older	. .	77·1
43	„	ten to fifteen years older	. .	66·6
17	„	fifteen to twenty-two years older	48·3

These tables are too limited to base any definite conclusions upon, as all statisticians will agree. Again, the number of children is not given, which is an unfortunate omission, as a percentage may be calculated from a very limited number. We are not informed definitely from what sources these tables were compiled, but if they are from official records it is a pity that the tables were not made much more extensive. For instance, the item which forms the most important test for my theory gives but *thirteen* fathers. The tables, however, bear the impress of truth and of scrupulous fidelity in their preparation, and yet I feel persuaded that if there is not a mistake somewhere, at least they are misleading as they stand. If viewed from the standpoint of the physical vigour theory, these tables would require us to believe that mothers of over forty years are physically so far superior to and more vigorous than their husbands under twenty-five, as to have more than two girls to one boy. Also that fathers of over sixty years are so far "superior," physically, to their wives under forty, as to have nearly two sons to each daughter.

Looking at the tables from my standpoint, the portion which troubled me most was that which shows a diminution of sons to fathers who were twenty-five to fifty years the seniors of their wives. I think a satisfactory explanation of this will be found if we take into consideration the motives which usually prompt the union of men of from fifty to seventy-five years and upwards with girls of twenty-five. It is, I think, obvious that few women in possession of such endowments as would warrant our classing them by my standard as " superior " would be likely to consent to unions with partners of three times their own ages. I consider, therefore, that the strong probabilities of such unions would be an excess of daughters. A 2 with B 2, A 5 with B 5, or even A 6 with B 6, would serve as examples of couples who would in spite of such a difference of age have an excess of daughters.

I have now, in an imperfect manner, sketched my progress from trembling hope to bold assurance, in the prosecution of my difficult task. I could not rest satisfied until able to tell approximately, upon seeing any couple, the proportion of the sexes of their children, and the *probabilities* when seeing but one of the parents. And this power I have now enjoyed for some years.

As I have stated a large number of difficulties, when discussing current theories in Chapter III., which those theories fail to explain, it may be well

to review them here shortly, with the light which
has now been thrown upon the whole subject, and
see what was the partial truth in each of them to
which it owed its currency. It will be seen by
reference to that chapter, that the difficulties, if not
already explained in this chapter, are yet obviously
and easily explicable on the theory I here put
forward.

It will be seen that my theory cannot be classed
as ovularian or spermatic, but that it supersedes all
theories based on any fixed "superiority" of this kind.
The mere act of volition by the mother, or letting
the imagination dwell upon one sex or the other, to
gain the desired end, is just as efficacious as to sit
idly and wish or imagine ourselves wealthy or wise ;
such notions, it is evident, have their origin in the
fact that they involve the most powerful impulses of
our nature. Thury's theory that the *time* of impregnation affects the sex may contain some grains of
truth so far as a natural process may be debilitatory,
as for instance in the case of delicate people, but
beyond this I can discover nothing in it.

The Alternate theory, as all must see at a glance,
is everything or nothing ; unfortunately, it is the
latter,—a mere surmise, in fact, and its authors do
not attempt to give it a scientific basis, and they
furnish no evidence of its truth. The Comparative
Vigour,—Relative Age,—or Nutritive theory, may
more correctly be termed a plain statement of
frequently observed facts, for the attempt made
to elucidate these phenomena displays nothing
approaching a scientific theory. Why it affords
frequent evidence of truth it would be superfluous

to repeat, as almost every chapter throws light upon the point. The theory of Genital Superiority stands or falls with the Physical Vigour theory; which, though seen in the light of my present theory to contain a glimpse of the truth, is quite incapable of accounting for the facts observed. The Right and Left theory was finally admitted by Dr. Trall to be false. The absurdities of Debay's theory have already been discussed. If his directions for obtaining *sex at will* were strictly followed—although he prescribes but for the mother—they would doubtless tend more or less to produce the desired result; for much that he recommends is incidentally conducive either to female "superiority" or "inferiority"; most of his directions as to diet, however, are in the main simply absurd. The Fowler theory is the least censurable of all—in fact, is quite commendable in the main. Fowler's idea of opposites cannot be improved upon, and it is a pity, with such an excellent hypothesis to start with, that he never found the means to verify it, and reduce his supposition to an absolute demonstration. Were my theory so impracticable that I could not *predict* sex with confidence, I should never rely upon it; and were it something so obscure or intricate that all might not likewise read and understand it, I should not care to present it to a busy world.

CHAPTER IX.

A UNIVERSAL BALANCE OF SEX THE PROOF OF MY THEORY.

IT has been insisted upon in the previous chapters, that, as the sexes remain substantially equal in numbers, a sufficient theory of sex must contain a principle which will hinder any great divergence by automatically producing a proportionate divergence in the opposite direction.

Since this Law of Sexual Balance must underlie our whole subject, it seems proper here to adduce some facts to prove that the principle of equalization, already set forth, is powerful enough to, and actually does, produce such a balance. I maintain that there is what physicists call a "stable equilibrium" between the sexes. A stable equilibrium Mr. Herbert Spencer defines as that operation by which "*any excess of one of the forces at work, itself generates, by the deviation it produces, certain counter forces that virtually out-balance it, and initiate opposite deviation.*"

Precisely the same thing obtains with regard to sex. The two sexes are in the aggregate equal in respect to temperament, so that very slight external causes may temporarily disturb the adjustment ; but the two great forces by their interplay will soon readjust the balance, if allowed to work unhindered. This, however, will only be true in the aggregate. In single families there will often be undoubted " superiority " on one side, and there will probably be an equal " superiority" in the offspring, though this may be modified by intermarriage with equally energetic partners of the opposite sex.

But in single cases where there is nearly an equality of temperament a very slight extraneous cause will determine the sex of offspring, and animals will be especially susceptible to slight influences from without, as they are not generally protected by civilization and its appliances from the operation of natural causes.

Thus it seems almost certain, to take an example, that meteorology, or the electrical condition of the atmosphere—to which males and females are not equally susceptible—affects the proportions of the sexes, slightly, every year; but a cold dry season giving an excess of males is sure to be succeeded by months that are warm and moist, and by an increase of females; and farmers and others notice this to be very often true of the proportion of sexes among their stock. This has been frequently observed to be the case in France. Certain it is that there are fluctuations of from 5 to 15 per cent. in the proportions of the sexes amongst domestic animals. The same meteorological conditions influence humanity, though of course to a less degree, owing to the power of civilization to ward off the effects of climate to a large extent; still the connection of epidemics with the weather and the perpetual variation of the death rate with the barometer prove unmistakably the influence of climate, even in civilized communities, and the popular mind has long ago recorded this in proverbs such as "A green yule makes a fat churchyard." The following is more specific, but points to the same conclusion. It is from a paper read by Colonel J. Puckle on "Meteorology in India in Relation to Cholera," in which attention was called to certain facts noted in connection with several severe outbreaks of cholera in the Mysore country during the last fifteen years:—

"On all these occasions there have prevailed abnormal meteorological conditions. Failure of the usual rainfall in the wet season, and an unnaturally high and moist temperature, have, as far as has been observed, been concurrent with the attacks in Mysore and Southern India. But notwithstanding the

study that has been expended upon the subject,
the clue to the mysterious origin of the disease is
not yet discovered, although the refuse that is
allowed to fester above ground is an active agent in
generating the malady."

Variations in the state of nutrition will also be
likely to produce greater effects with the lower
animals than with man, as the former are unable
to supplement the actual food given them by stimulants, or now and then by small portions of better
food, which in the case of human beings render the
results less definite.

M. Charles Girou took a flock of some 270 ewes,
and divided it equally, with the avowed purpose of
producing a large excess of males in one division,
and of females in the other. He accomplished his
end, and the result, as a whole, is confirmatory of
my theory, but some of its details, as may be seen,
were somewhat perplexing.

Age of Mothers.	Sex of Lambs. Males.	Females.	Age of Mothers.	Sex of Lambs. Males.	Females.
2 years	14	26	2 years	7	3
3 years	16	29	3 years	15	14
4 years	5	21	4 years	33	14
Total	35	76	Total	55	31
Five years and older, 18		8	Five years and older 25		24
Total	53	84	Total	80	55

There were two rams with this flock; one fifteen months and the other nearly two years old. Rams and ewes were all placed in the richest of pasturage—were highly fed.

Two strong rams, one four and the other five years old, were with this flock, and all were kept on a scanty supply of food.

Girou, and all who hold to the theory that the
physically "superior" originates the sex, find in the
above experiment the following proofs :—

Young, immature rams are placed with ewes in the best possible physical condition, owing to rich and abundant food, and the result is an excess of over 50 per cent. of females, due to maternal *physical* superiority. Again, mature old rams are placed with an ill-assorted flock of ewes in a poor pasture, and the consequence is nearly a 50 per cent. male excess, owing to masculine *physical* " superiority," or feminine " inferiority," as one may prefer to call it. But I view the case in a very different light: my version of it is the following : Two young rams that have gained their normal weight, and that are in fact in the very prime of their existence, and sufficiently well-nurtured to meet almost any drain upon life's fund, are placed among a flock of sheep, dull and inactive from over-feeding, and therefore " inferior ; " the young rams are not deteriorated in this way, as their food is only sufficient for them, owing to the heavy drain upon their reproductive system. Again, two rams past their prime are turned in with sheep that have to work for a living, that is, to wander about the pasture in search of food, which elevates them by stimulating them to activity. The rams have not only to search for food, but also to meet the drafts inseparable from the circumstances, and thus perhaps deteriorate somewhat, while the ewes are kept in the most favourable condition for " superiority ; " that is, with the nervous system in the ascendant. The total result is not different from what I should have predicted.

We see, then, that animals are distinctly affected by nutrition, and we cannot avoid the conclusion that men must also feel something of the same influence; indeed, it is known that in Belgium even the price of bread, as already mentioned, slightly affects the proportion of the sexes.*

The point here sought to be brought out is that very slight causes may influence temporarily and locally the proportion of the sexes, owing to the near temperamental equality of the aggregates.

The working of the law of equilibrium as operating in the case of animals, to restore the balance thus disturbed, may be further illustrated. Thus Darwin says: " The fluctuations in the proportions during successive years (among animals), are closely like those which occur with mankind when a small and thinly populated area is considered: thus in 1856 the male horses were as 107·1, and in 1867 as only 92·6 to 100 females. In the tabulated returns the proportions vary in cycles, for the males exceeded the females during six successive years; and the females exceeded the males during two periods, each of four years: this, however, may be accidental; at least, I can detect nothing of the kind with man in the decennial table in the Registrar's Report for 1866."

The record of the above-mentioned racehorses for 21 years gives 99·7 males to 100 females, which surely is not far from a perfect poise. Considering the length of these cycles, it seems to me most likely that the result is governed by other than climatic influences. My idea of its probable cause may be thus expressed: Suppose a dozen choice stallions are introduced into any district; their offspring will

<hr>

* Supra, Chapter II., p. 22.

number from one to two thousand each year, and, owing to their pedigree, fillies will naturally preponderate—especially during the first few years. These young mares, in their turn, being of a choice strain, will, in a majority of instances, be "superior" to the sires employed, and *colts* will predominate in their issue.

Thus does "superiority" pass from one sex over to the other, through each successive generation, interwoven and ever crossing, yet producing a symmetrical whole; in other words, the great sexual balancing force is ever moving backwards and forwards like the flood and ebb tides, over every country of the world.

If there be an excess of males from any cause in one generation, this is due to superior energy and activity in the mothers; the males in excess will inherit this "superiority," and in their turn will have an equal excess of female offspring, and so on. This is equally and more strikingly true of single families, as I shall presently show, for a family is more frequently isolated, and thus observation is more likely to lead to accurate results. It is notorious, as we have already seen, that when there is an excess of males in one generation, there will be a compensating excess of females in the next, and *vice versâ*. This is brought about in two ways. Not only will males, above the average standard, have an excess of daughters, but females, below the average, will on their marriage tend to have daughters, even by husbands who are only of the average standard, and *a fortiori* by those above it; and the converse will be the case in the next generation. This is of course only a general tendency, and will show best in large numbers, for allowance is always necessary for the relative " superiority " of those with whom the members of the special family intermarry.

Here let me make myself perfectly clear as to this principle of adjustment passing from generation to generation. I do not mean to assert that there is a principle of adjustment apart from, and independent of, the relative " superiority " of individual parents, but simply that the operation of heredity *tends* to prevent any preponderance of one sex *ultimately* resulting from the union of two persons very unequal in a sexual point of view. If the father, in any family, is very " superior " to the mother, and there are consequently five daughters to one son, these daughters will be likely to show in their character, temperament, features, etc., a strong resemblance to their father, and as a natural result they will probably be " superior " to their husbands, who will presumably represent, on the whole, *men of average type*, while the wives will be *above* the average of women. If this be the case, the result will be families with an excess of boys, in consequence of the individually inherited " superiority " of the respective mothers ; and this excess will correct the feminine preponderance of the preceding generation. These sons again inheriting, in a somewhat less degree, their mother's "superiority," will probably be the " superiors " of their wives, and in the next generation girls will be found slightly in excess ; so that the total of the three generations will be an almost equal number of both sexes, though each generation taken alone shows a preponderance of one sex. This is the case where there is at the commencement inequality, but if the parents are really equal at first, there will be an equal distribution of the sexes in their family, and no readjustment will be necessary; on the other hand, for every degree of inequality of the sexes in a family there will be a tendency, *pro tanto*, for the families of the children to produce a deviation in the other direction, and so to restore the numerical balance. Even if daughter's inheriting their father's "superiority" should *all* marry men of equal "superiority" (a most improbable event), they would still have families with an equal number of sons and daughters; for both parents being "superior" to the average, the tendency would be to equality in the sexes of their offspring; therefore even in this extreme case the original deviation would be arrested in the second generation.

From what has been said it will be seen that the individual "superiority" as the *determining factor* is not overruled at all, but simply explained in some cases. The law of heredity merely *accounts for* the "superiority" manifested in the children. The daughters of an able man and the sons of a gifted woman will exhibit the same signs of "superiority" as were shown by their parent, subject to some modification from the other parent, and perhaps from atavism. I should therefore say that the pedigrees of any couple if presenting any special features would aid us in predicting the sex of their offspring, and would be a help also in interpreting the signs of "superiority" in their own character and lineaments. The principle of readjustment by heredity is there-

fore a *consequence*, and not a contradiction, of the law
above stated of the determination of sex by the
"superior opposite" parent. Every case, however,
must be judged upon its merits, as the law of
heredity is uncertain in its operation and little
understood, while on the other hand there is an
important new factor to be taken into account with
each succeeding generation, viz., the temperament
and influence of those with whom the children intermarry. Nevertheless the adjustment as stated above
will appear plainly enough in most cases, and will
certainly be found to operate if the family be followed to the second and third generation,—and it is
to aggregates that we must look for the demonstration
of a *general law*.

It may possibly make my argument clearer if we
notice how this principle of adjustment is lacking in
current theories. We will take, for example, the
physical vigour theory, according to which a man of
superior physical vigour will have a family with
sons in excess. In the next generation these sons,
inheriting their father's vigour, will all tend to have
an excess of sons in their families. In succeeding
generations this male excess would be increased, and
at last the preponderance of males would be enormous. This result we know does not ensue, for
the sexes are very nearly equal, not only throughout
the world, but in each family in the second and third
generation.

There is another way in which the adjustment of
the sexes may be brought about, and that is by what
I may call duplicate inheritance. If we suppose a
"superior" father represented by 125, and a mother

by 75, their children would of course, on our theory,
be chiefly girls inheriting a portion of their mother's
with a great deal of their father's qualities, and
though probably above the average, they would be
less so than their father, ranging possibly between
105 and 115. Consequently, supposing them married
to average men, their families, though showing an
excess of sons, would not present so great a disparity
of the sexes as was seen in the previous generation.
In this perfectly simple manner Nature brings about
a readjustment of the equilibrium of the sexes,
neutralizing the disturbance of her balance by the
marriage of two persons, sexually, very unequal.

A similar readjustment will follow the disturbance
produced by disease. It is well known that many
epidemics—for instance, cholera, as already referred
to in Chapter II.—prey excessively upon the male
population. The men then will be left in a minority.
But among the consequent excess of women, *selection* now naturally comes into play, and only the
"superior" ones are married, or rather the "inferior"
and less attractive women will fail in securing
husbands, because of the minority of the males.
The result of this selection of the best women as
wives must inevitably be an excess in the proportion
of sons born, the excess bearing a direct ratio to the
previous mortality among the men, and the consequent severity of the process of selection among
the redundant women. The same will be seen after
any derangement of the numerical balance by excessive drafts of men for war, as in France under
Napoleon, cited in Chapter II. The best men,
physically, are in this case selected and killed,

leaving the "inferior" members of the male sex to perpetuate the race. These will in a majority of cases be "inferior" in all senses, and of course in the special sense in which I use the word, to their wives, and there will be in the succeeding generation an excess of sons. This harmonizes with the facts, for by the year 1830, *i.e.*, in one generation, nearly all traces of the ravages of war were obliterated from the census returns in France. This readjustment applies, however, to numerical balance only. It is said that the average stature of men has been lowered in France from two to three inches in consequence of those wars.

Another application of the theory would explain the statistics of Australia, also noticed in Chapter II. Here we see an immense excess of males living in isolation, and therefore in enforced continence, while a hard life developes their energies to the utmost. The "superior" will be the only ones likely to find wives, and among such unions we may confidently look for an excess of daughters, which will tend to make up the deficiency.

The same theory also explains the difference in the proportion of the sexes between different countries, between urban and rural populations, between legitimate and illegitimate births, and between polygamous and monogamous unions. It also throws light upon the curious difference between Jewish and Christian populations, living side by side in the same country; full explanations on these points will presently be given.

My theory that this reciprocal adjustment is accomplished by the "superior" parent determining for the opposite sex is supported by the law of transmission of resemblances, concerning which there is no difference of opinion. In a general way this law is expressed by Dr. Hough as follows :

" 1. Children resemble their mothers more than their fathers.

" 2. Males resemble their mothers, and females their fathers.

" 3. When children do not resemble their parents, but their grandparents, males resemble their maternal grandfather, and females their paternal grandmother."

This is precisely the truth of the matter ; and for the reason that a daughter owes her sex to the fact that her father contributed more to her existence at the time of conception than the mother, boys owe their sex to their mothers having more of the transmissible essence of life to give. For the most part, this is strongly corroborative of the truth of my theory, and in further confirmation I will cite a variety of evidence. No witness could be more impartial than M. Girou, who gives the following table :—

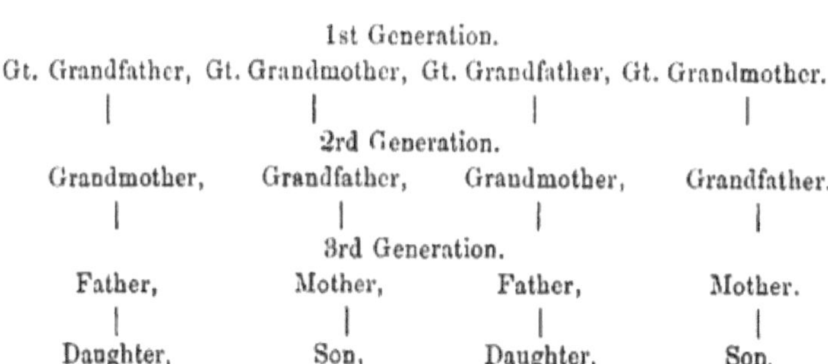

1st Generation.

Gt. Grandfather, Gt. Grandmother, Gt. Grandfather, Gt. Grandmother.

2rd Generation.

Grandmother,   Grandfather,   Grandmother,   Grandfather.

3rd Generation.

Father,   Mother,   Father,   Mother.

Daughter,   Son,   Daughter,   Son.

From this we gather that, first, the daughter tends to resemble the father ; second, as he again resembles his mother they must both resemble his mother ; third, that the daughter chiefly resembles her father's mother's father—for there is a blending of two natures in each generation. Even Aristotle noted this principle of opposite resemblances. Sportsmen have long adopted the proverb, "*Chien de chienne et chienne de chien*," to express the same thing among their hounds—the qualities of the mother appearing in the "bull pups," and the father's in the "belles." Buffon notes the same, as also does Lucretius :

> " Thus of the father's likeness does prevail
> In females, and the mother's in the male."

The testimony of the *National Live Stock Journal*, quoted in Chapter III., may also be referred to as establishing this point.

Most famous men have owed their " superiority " to their mothers, and this is commonly expressed in the saying that " great men have great mothers " ; while, on the other hand, the women known as celebrities reflect the genius of their fathers : Sarah Castor, the mother of John Milton, was a woman of incomparable virtue and goodness The mother of Napoleon the Great was noted for the qualities her son displayed ; the mother of Byron for several of her son's leading characteristics. Marcus Aurelius inherited the virtues of his mother, and Commodus the vices of his Charlemagne shut his eyes to the faults of his daughters, because they recalled his own. Cornelia, the mother of the Gracchi, was a daughter of Scipio. Arete, Hypatia, Madame de Staël, and "George Sand," all had philosophers for their fathers. The mother of Tasso had the gift of poetry. Burns, "Rare Ben Jonson," Goethe, Scott, and Lamartine, were all the sons of women remarkable for their "vivacity and brilliancy of language." M. Bernard Moulin, speaking of the influence of parentage, gives the following: "History shows to us the parents of Raffaelle both devoted to the art of painting. The wife, a true Madonna, delighted in subjects where grace and piety prevailed; the husband, a great dauber, preferred strength for his part." It is unnecessary to point out that the son inherited his *mother's* tastes far more than those of his father. The genius of Bacon, Buffon, Condorcet, Cuvier, D'Alembert, Forbes, Gregory, Watt, Brodie, Jussieu, and Newton is plainly traceable to the maternal side. When it can be shown that some distinguished woman had a clever mother, she is doubly fortunate, as her father must have been the mother's equal or "superior" in some important respect: the converse of this is true of famous men.

From the foregoing we should expect to find that a collated list of the ancestry of great men would bring out this point with great force, and be a striking confirmation of my theory. Mr. Galton's work *Hereditary Genius* seems at first sight to offer the very conditions desired, and his testimony is at once impartial and unimpeachable. He, however, seeks only for the sons and other male relatives of distinguished men, having no idea of the connection here established between the sexes in respect to inheritance. Consequently, though the value of his evidence for my purpose is enhanced so far as it goes,—he being unintentionally hostile to my theory,—he has in the majority of cases said but very little about the mothers, wives, and daughters of the representative men of genius whom he has selected. But in looking carefully through his list of some hundreds of statesmen, warriors, literati, men of science, and poets, *in every instance where the mother is incidentally mentioned she is described as a woman of remarkable ability or intelligence.*

As these instances tell without exception so strongly in favour of my theory, I will here give them in detail.

STATESMEN.—*Canning's* mother had great beauty and accomplishments.

William Pitt the younger. His mother was Lady Hester Temple, of a remarkably gifted family.

Sheridan's mother, Frances Chamberlain, wrote comedies which won the approval of Garrick, and in spite of her defective education her novels were praised by Fox and Lord North. Her father, the Rev. Dr. P. Chamberlain, was an admired preacher, but also a humorist, and full of crotchets. Sheridan, himself an oratorical genius, married a musician, Miss Linley; and had a son Thomas, whose family was large, and whose daughters were especially able. One became the Hon. Mrs. Norton, the poetess, and the other the mother of the late Lord Dufferin.

Sir Anthony Cooke and Sir T. More I shall mention more particularly farther on.

WARRIORS.—*Alexander the Great.* His mother,

Olympias, was ardent in her enthusiasms, ungovernable in her passions, ever scheming and intriguing; she suffered death like a heroine. Of Alexander, Galton says that he had four wives, but only one son, making no mention of daughters, though I should have expected to find he had an excess of daughters. This is probably simply because the daughters were not reckoned.

Fitzjames, Duke of Berwick. The illegitimate son of James II. by Arabella Churchill, sister of the Duke of Marlborough.

Napoleon. His mother, Letitia Namolini, was a heroine by nature, and one of the most beautiful women of her day. Galton is loud in her praise, as firm and energetic, *i.e.* "superior." His father was a Corsican judge of no special ability.

Julius Cæsar. His mother, Aurelia, "appears to have been no ordinary woman."

Gustavus Adolphus. Queen Christina was his only child. She was "a woman of high abilities, but of masculine habits." Gustavus Vasa, his ancestor, had also a remarkably clever daughter, Cecilia.

P. Cornelius Scipio had a daughter Cornelia, "the mother of the Gracchi." She "was almost idolised by the people," and was one of the greatest women of antiquity.

Saxe. His mother was "beautiful but frail." A natural daughter of his was grandmother to Madame Dudevant (George Sand), the novelist.

MEN OF LETTERS.—*Hallam.* His mother was the daughter of R. Roberts, M.D., and was a lady of ability, somewhat over anxious. She resembled her son in features, but had only two children that lived.

She and her husband must have been a well-matched couple, having had a son and a daughter, but both being cerebrally "superior," they were not prolific.

Richard Porson. His mother was a housemaid in a clergyman's family. She read her master's books in secret, and was found to possess a considerable knowledge of Shakespeare and other authors.

Anthony Trollope. His mother was a celebrated novelist.

MEN OF SCIENCE.—*Bacon's* mother, Anne Cooke, was one of a family of five *sisters*, all gifted, and was " herself a scholar of no mean order."

Hon. R. Boyle. His mother Catherine is spoken of as "religious, virtuous, loving." He had four *brothers* and seven *sisters*, the latter of " considerable merits."

Buffon said he derived his qualities from his mother.

Cuvier. His mother was an accomplished woman, who took especial care with his early education. He had one *brother.*

D'Alembert (illegitimate). His mother, Mdlle. de Tencin, was a novelist of high ability, and she had two *sisters* of note.

Darwin. His mother was Josiah Wedgwood's daughter. (His grandfather was Erasmus Darwin.)

Forbes, E. His mother was gentle and pious, and passionately fond of flowers; she had *three other sons.* E. Forbes himself had no issue.

Watt. His mother, Agnes Muirhead, a superior woman, is thus described by one who knew her: " A braw, braw woman ; none now to be seen like her."

Poets.—*Byron* had a mother of strong character, and a daughter, Ada, with remarkable gifts.

Goethe. His mother was remarkably clever, and his father cold and pedantic.

Tasso. Porzia di Rossi, his mother, was a "superior" woman in every respect.

In further illustration of this subject it will be well to quote the following from Mr. Galton's *Hereditary Genius :—*

" I was desirous of obtaining facts bearing on heredity from China, for there the system of examination is notoriously strict and far-reaching, and boys of promise are sure to be passed on from step to step, until they have reached the highest level of which they are capable. The first honour of the year in a population of some 400 millions, the senior classic and senior wrangler rolled into one, is the 'Chuan-Yuan.' Are the Chuans-Yuans ever related together? I put a question on the subject into the pages of the Hong-Kong *Notes and Queries* (August 1868), and found at all events one case, of a woman who, after bearing a child who afterwards became a Chuan-Yuan, was divorced from her husband, but marrying again, she bore a second child, who also became a Chuan-Yuan, to her next husband. I feel the utmost confidence that if the question were thoroughly gone into by a really competent person, China would afford a perfect treasury of facts bearing on heredity."

And finally, I may call attention to Mr. Galton's chapter on "Comparison of Results," in which indirectly and quite unintentionally he draws several conclusions which are strongly in favour of my theory. It is to be noted also that he finds singularly few eminent fathers of eminent men, though his wish and endeavour have been to discover as many as possible. Had he made equally careful search for superior *mothers* there would have been no dearth, as his own statistics above quoted clearly show, but family and other records are unfortunately too often wanting in the very particulars which are required to establish this fact. Moreover the ability or superiority of a woman is very likely to be overlooked or unrecognised, as there are fewer opportunities for women to make their talents known. Mr. Galton makes the following remarks in the chapter above mentioned :—

" In contrasting the columns B of the different groups, the first notable peculiarity that catches the eye is the small number of the sons of commanders ; they being thirty-one, while the average of all the groups is forty-eight. . . . Though the great commanders have but few immediate descendants, yet the number of their eminent grandsons is as great as in the other groups. I ascribe this to the superiority of their breed, which ensures eminence to an unusually large proportion of their kinsmen.

" The next exceptional entry in the table is, the numbers of eminent fathers of the great scientific men as compared with that of their sons, there being only twenty-six of the former to sixty of the latter, whereas the average of all the groups gives thirty-one and forty-eight. I have already attempted to account for this by showing, first, that scientific men owe much to the training and to the blood of their mothers

" The next peculiarity in the table is, the small number of eminent fathers in the group of poets. This group is too small to make me attach much importance to the deviation; it may be mere accident.

" The artists are not a much larger group than the poets, consisting as they do of only twenty-eight families, but the number of their eminent sons is enormous and quite exceptional."

These results are all the more striking as the tendency naturally is for able women to unite with able men, and thus one would think them likely to be in most cases " inferior" to their husbands, though far above the average of their sex; the presence, therefore, of a large number of sons is quite in accordance with my theory. In order to show that this is no imaginary conclusion to account for facts, I may once more cite Mr. Galton, who says :—

" The large number of eminent descendants from illustrious men must not be looked upon as expressing the results of their marriage with mediocre women, for the average ability of the wives of such men is above mediocrity. This is my strong conviction, after reading very many biographies, although it clashes with a commonly expressed opinion that clever men marry silly women. It is not easy to prove my point without a considerable mass of quotations to show the estimation in which the wives of a large body of illustrious men were held by their intimate friends, but the two following arguments are not without weight. First, the lady whom a man marries is very commonly one whom he has often met in the society of his own friends, and, therefore, not likely to be a silly woman. She is also usually related to some of them, and therefore has a probability of being hereditarily gifted. Secondly, as a matter of fact, a large number of eminent men marry eminent women. If the reader runs his eye through my Appendices, he will find many such instances. Philip II. of Macedon and Olympias; Cæsar's liaison with Cleopatra; Marlborough and his most able wife; Helvetius married a charming lady, whose hand was also sought by both Franklin and Turgot; August Wilhelm von Schlegel was heart and soul devoted to Madame de Staël; Necker's wife was a blue-stocking of the purest hue; Robert Stephens, the learned printer, had Petronella for his wife; the Lord Keeper, Sir Nicholas Bacon, and the great Lord Burleigh, married two of the highly accomplished daughters of Sir Anthony Cooke. Every one of these names, which I have taken from the Appendices to my chapters on Commanders, Statesmen, and Literary Men, are those of decidedly eminent women. . . . The great fact remains, that able men take pleasure in the society of intelligent women, and, if they can find such as would in other respects be suitable, they will marry them in preference to mediocrities.

"I think, therefore, that the results given in my tables under the head of 'sons' should be ascribed to the marriages of men of Class F and above, with women whose natural gifts are, on the average, not inferior to those of Class B, and possibly between B and C."

But if the view of the case above stated be true, viz., that the abilities of the parents pass to the child,

the converse is equally true, though far less pleasant
to dwell upon: the evil tendencies, unbridled passions, and criminal or drunken habits of parents are
equally certain to be transmitted to their offspring.
Habitual intoxication in the parents only too frequently produces insanity, idiotcy, or a predisposition
to excessive drinking in the child, and so with various
other forms of depravity.

I can call to mind but three conditions under
which the law of Balance would be almost powerless
to establish an equality: 1, When emigration and
kindred causes constantly drain a percentage of one
sex from a country—as in Great Britain, where,
according to the Registrar General's reports, out of
every one hundred females over twenty years of age,
fifty-seven are wives, thirteen widows, and thirty
spinsters; 2, When the infanticide of one sex is
practised—as amongst the aborigines, in some parts
of Australia, where the men are in excess as *three*
to *one*, owing to the destruction of female infants;
3, When there is uniformly a great disparity in the
ages of parents equally moral; in many countries
the husbands are probably on the average ten years
older than their wives. The first and second cases
are out of the province of nature to adjust, as her
sexual laws can only act through the reproductive
function; she pays no attention to migration or
to murders. Regarding the third case, I cannot but
think that a difference in years is unnatural, and must
be unfavourable in some way to offspring, although
generations may be required to develop it; but of
this I have no conclusive statistical proof; nevertheless it is certain that this inequality of the parents

does cause a distinct or permanent disproportion of the sexes, which, however, is not real, but only numerical. This may partly account for the excess of male births in civilised countries, where the husband is usually older than the wife.

In the various provinces of India the women number only from 82—93 to 100 males : accounts seem to ascribe similar habits of virtue and vice to both sexes; brides, however, are remarkably young there, which probably explains the excess of males. The practice of infanticide has also induced a habit of carelessness as regards daughters. In Bengal, the proportions are more nearly equal, yet while there are nineteen boys to sixteen girls, there are thirty-four women to thirty-one men. Some allowance must be made on these latter figures, as girls are often prematurely reckoned as women.

There is a notable difference in the respective ages of Jewish couples : Prussian statistics—the most reliable of all—show that the excess of male births among them, in different districts, ranges from 112—144 to 100 female, and at the same time there are fewer men than women : migration and morality are doubtless factors in this problem. While among the Christians there, the adult men are as 100 to 100·72 women, the Jews are only as 100 to 103·37 women.

In Montevideo, with a population of 100,000, the sexes are born in the proportion of *eleven males* to *seven females*, and yet there exists a lamentable scarcity of marriageable young men. The difference in age of those marrying averages nearly ten years, which in part causes this disproportion, but masculine immorality does far more. Deaths within a week after birth are here alarmingly frequent.

In Philadelphia there are 110·44 males born to 100 females : there is no permanent excess of males, however, but the reverse ; 113·44 male deaths occur to 100 female, and the population of the city at large gives but 89·5 males to 100 females. In London, for many years past, males have been born in excess, though in the census returns they are only found to be as 85 to 100 females. In almost every European country there is at least an excess of 4 per cent. of male births, but the mortality among males is from 3 to 4 per cent. more than among females, making women proportionately redundant.

By the United States Census, we see that substantially the same obtains, yet with the coloured population this is all reversed ; females are born and die in excess among them, by over 3 per cent.

It may occasion some surprise that the difference in the ages of parents can prove so fatal to the offspring when all around us we find thrifty, promising children—chiefly boys—with fathers a quarter of a century older than their mothers ; but these cases are exceptional. The fathers are usually *moral* men, living with second wives ; in countries where we do not find this to be the case the average difference in years is over ten.

Those who, at thirty, marry wives of twenty, have more or less in all countries, not only an extra decade to tell against them in the sex of their children, but in many cases have also a weakened constitution resulting from past excesses or dissipation. Vienna, notorious for a laxity of morals, had in 1819, notwithstanding the large excess of male births, an adult population of two women to every man.

F. R. Burton, in his work "The Highlands of Brazil," gives the following figures from the census returns of 1859 for the town of Sao Joao d'El Rei :—

Men free .	3150
Women free	4650
Men servile .	260
Women servile	390
Foreigners .	50

That is, an excess of nearly 50 per cent. of women as compared with men, or three women to two men. This is what might be expected from the large intermixture of the white race with the coloured women.

Again, Mr. Williams states that in the Pará, births are in the proportion of four or five females to one male. Colonel Tulloch, in the *Anthropological Review*, says that the negroes in the English Antilles will be nearly extinct within a century, and Mr. Burton says the decrease amounts to four per 1000 per annum.

The above facts are in perfect accordance with my theory. Nature weighs the parents against each other, and any defect of constitution, either from immorality or from age, must inevitably be made manifest ; the sex of the offspring will be that of the parent found wanting when thus weighed. Man is more profligate than woman. and therefore by the law of opposites there is an excess of male offspring, the mothers being " superior " in a majority of cases. Thus the proportion of the sexes born in any country is an excellent criterion of its comparative morality. Nature, by producing an excess of males, publicly proclaims that men are almost everywhere more given to vice and dissipation than women—the surplus quantity being born to compensate for the defective quality. Or the case may be stated thus: As a general proposition, families with an excess of girls have good fathers and mothers, and are consequently healthy; those with a surplus of boys frequently have bad fathers,—hence arises the mortality they exhibit. In other words, girls generally have good mothers, while many boys have bad fathers.

Quetelet, the Belgian statistician, finds that women live longer in cities, and men longer in the country districts; also that in cities male births are in an excess of *one per cent.*, a similar excess of female births occurring in the country. This latter fact is presented as one of the " curiosities of statistics," but in the light of this volume it can only be classed as *interesting*. One of these facts is the sequel to the other: women in cities live a superior life to those in the country, as is proved by their lengthened days, and this will account for the slight increase in the proportion of boys: in the country, the reverse of this produces an opposite result.

Darwin speaks of those " mysterious influences " which uniformly produce a larger proportion of females among illegitimate than legitimate offspring; yet in prostitution a large excess of male births appears. The " mystery " vanishes when we apply the theory of *superior opposites*. The fathers of most illegitimate children are by no means " dissipated" profligates, but are refined and seductive in manner,—to which fact, in many cases, they owe their success in deluding their victims.

I have studied numerous instances of persons of a superior race intermarrying with an inferior, the result always being in perfect accord with my theory of sex. In the case of the mulattoes already mentioned, from white fathers and black mothers, instead of their showing like the whites, a 5 per cent. excess of male births, they show from 12 to 15 per cent. of female redundance. If these white fathers were less dissolute in their habits, there would be a still larger proportion of female births. On the other hand, were the colours reversed—white mothers and black fathers—there would undoubtedly be a striking surplus of male births. This excess of female mulattoes, together with the remarkable instance of the Lipplappen, who, as already mentioned, become sterile after the third generation when crossed with a superior race, and other similar instances which I have cited, I consider the most striking confirmations possible of my theory.

As regards the comparative morality of the sexes throughout the world, women appear, almost everywhere, in the more favourable light, and as a matter of course more males are born. According to the *London Review*, the criminal statistics in different parts of Great Britain show that seven men to one woman are found guilty; and of juveniles, four boys to one girl. In the " Rogues' Gallery," at the New York Central Police Station, there are 1,274 male faces, and but 50 female. In one of the State

Prisons in New England I notice 237 males to 10 females. All this goes to prove the comparative morality of the sexes.

Jewish women are more chaste than any others, and the proportion of males born to them is enormous; illegitimate children are very rare amongst them. The licentiousness of the Jews in early times is well known, but it appears that polygamy was abandoned after the Babylonish captivity. For centuries there has been a minimum of crime amongst them, and illegitimacy is looked upon as a lasting and deep disgrace.

Being aware of the excessive mortality among male children in the United States—even in still births 50 per cent. more—and attributing it largely to paternal frailties, I thought that idiots of the male sex must greatly preponderate. On investigation I found, according to the Ninth United States Census, that there was an excess of 40 *per cent.* of male idiots. With the insane it appears, as might be anticipated, that woman is rather more frequently driven to insanity than man—6 per cent. difference. In regard to the blind, we find an excess of 27 per cent. of males; and of the deaf and dumb 23 per cent. These figures of themselves could never establish any theory, but in connection with so many other facts, all pointing in the same direction, they certainly tend in my favour.

Ever since statistics have shown that in the more enlightened countries of the world about twenty-one males are born to twenty females—5 per cent. excess—it has been said that this is Nature's wise provision for supplying deficiencies caused by accidents and war, man's life being so much more exposed than woman's. But the tables show that the number of men who really die thus is only about one per cent. more, when offset by the casualties to which the other sex is exclusively exposed—such as the perils of maternity. It is true that *seven* men die suddenly to *one* woman, and that they die unnatural deaths from murder, suicide, drunkenness, etc., at the rate of *five* to *one*, all of which leads to an increase of male births—owing to masculine degenerateness. Man by his excesses undermines his constitution, and thus lowers his "superiority"—in the aggregate I mean; the result is a surplus of male births, by which nature restores the balance, as the weakened males will die on an average earlier than the females, and of this every insurance or annuity table will furnish proofs. That this male increase is due to masculine immorality is further shown by the larger excess of male births and male infant mortality in France than elsewhere, and by the larger excess of male births among urban populations, as proved by the various quotations given in Chapter II. Nature never makes the sex of a single child dependent upon the ravages of war or accident. It is ridiculous to suppose that because twenty men, out, for instance, on an excursion, meet their death by accident, Nature stops her ordinary processes and begins moulding sex to meet the emergency. Nature stands blindfolded with scales in hand, like Justice, uninfluenced by wars, politics, nationality, religion, or even death itself; and impartially awards sex according to the merits of each individual case. Mr. Greg, in discussing the question, " Why are women redundant ? " after calling attention to the excess of 4 or 5 per cent. of eligible husbandless ones, pertinently asks, whether Nature designs these for celibacy or polygamy. Nature really intends neither. It is best for every man to have a wife and every woman a husband, which would also be quite possible were men but disposed to live as virtuously as women do. Innocent maidens are condemned to lead lives of hopeless celibacy because of the immoralities of the sterner sex.

In further support of this assertion, and in illustration of the working of the principle of adjustment, I will now try and present proofs drawn from the pages of history as well as from biographies, family genealogies, etc. These proofs are, incidentally, so many further corroborations of the truth of my theory as applied to families ; in this sense they are supplementary to the verifications given in the last chapter. The tedium involved in obtaining these facts is almost incredible : one may trace the marks of greatness, or " superiority," in the life of some great man, hour after hour, growing impatient meanwhile to discover the sex of his children, and at last only learn that " he never married," or " he left no issue," or possibly that " five children survived to mourn his loss," but not a word as to their sex ; and even if the sex be stated, the information is, in most cases, far from complete, for we are told nothing of the character of the wife, and we can never judge accurately if acquainted with but one of the parents. With the exception of M. Girou, I know of no one likely to sympathize with me in these vexatious labours. His research through records spread over more than eight centuries of the most eventful period of French history has satisfied me that there is little occasion for me to go over the same ground. The fact that he sacrificed so much time to the study of this sexual problem, and, while entertaining the hypothesis that he did, only obtained results so unwelcome to him, affords at once the strongest possible proof of their correctness, as well as decisive confirmation of the truth of my theory. He finds, as previously stated,* that *all men of remarkable force of character were the parents of daughters chiefly, and that women of similar traits had chiefly sons.* If written to order, these conclusions could not be better adapted to my purpose; and I consider it fortunate that they come from such an independent source.

I will now cite a few illustrations, enough to show clearly the general indications of "superiority," and so to impress them upon the reader that to judge accurately of a person may be as easy and intuitive an act as it now is to estimate any one's age. For my own part, I have seen my theory so repeatedly verified that I can feel little interest in further search.

In searching for "superiority," it is ever well to bear in mind the declaration of Shakespeare, that "some are born great, some achieve greatness, and some have greatness thrust upon them." *Fame* is not the best criterion of intrinsic force of character —of "superiority."

Mary, Queen of Scots, gave birth to King James I.

* Chapter VIII.

at the age of *twenty-four*. Her husband was only *twenty*, and is described by Keith as tall, handsome, and possessed of most of the fashionable accomplishments of his day, all of which speaks well for a daughter—*i.e.* for his "superiority;" but mark what follows: "He was addicted to intemperance, to base and unmanly pleasures, and was so very weak in mind as to be a prey to all who came near him; inconstant, unable to abide by any resolution." Such being his character, volumes could add nothing to the strength of this case.

Mr. Swinborn, in speaking of the Empress Maria Theresa, remarks: "She has such an internal fever and heat of the blood, that she cannot bear to have the windows closed at any season of the year." This is but a partial glimpse, yet it reveals a great deal. Any excess of one function or faculty is, with the rarest exceptions, accompanied by deficiencies equally apparent. This description indicates that the Empress was of the sanguine temperament, that the nervous system did not predominate in her, but the fleshy, the physical. She was married at the age of eighteen; her first-born were twin daughters, and in twenty years she had *twelve* daughters and *four* sons. I find conflicting statements regarding her husband, but as long as the above quotation stands unimpeached, the case I hold to be in my favour.

Sir Anthony Cooke was "a man of rare talents and of a reflective turn of mind;" he had *one* son and *five daughters*. "His daughter Anne was the most gifted." She married Nicholas Bacon, eighteen years her senior. They had two sons, Anthony and Francis Bacon; the mother was thirty-two at the birth of Francis. With such a pedigree, need we wonder at the talents or sex of the offspring?

The superior qualities of Sarah Castor, the mother of John Milton, are well known. Milton's wife was not possessed of more than average abilities; they had *three daughters*. Only one of these had children. She married a weaver, and bore him *seven* sons and *three* daughters, which shows that she was by inheritance "superior" to her husband. Milton's family is a perfect illustration of the law of sex.

Madame Guizot, mother of the famous French statesman and historian, "was a woman of marked talent," and of a family long distinguished for brilliancy: she had *two* sons.

Of the mother of the three Wesleys, her biographer says: "She was an admirable woman, of highly improved mind, and of a strong and masculine understanding: an obedient wife, an exemplary mother, and a fervent Christian."

I think no one will dispute that the last two I have named were "superior" women. As little or nothing is said of their husbands, we are left to infer that they were merely men of ordinary intelligence. Had these ladies become the mothers of as many daughters as sons, this would have indicated "superior" husbands—not superior to their wives, but to the average of mankind. I cannot imagine such women having daughters only.

The stern Reformer, John Knox, had daughters exclusively. The Rev. John Welch—who at one time led a profligate life—married one of these daughters and had three *sons* by her.

Sir Thomas More, so distinguished in the legal profession, and who suffered death rather than do violence to his conscience and religious convictions, married at twenty-one "a country girl, very ill-educated, but fair and well formed." Erasmus says of the marriage : "He wedded a young girl of respectable family, but who had hitherto lived in the country with her parents and sisters, and who was so uneducated that he could mould her to his own tastes and manners. He caused her to be instructed in letters, and she became a very skilful musician, which particularly pleased him. The union was a happy one, but short, the wife dying, and leaving behind her a son and *three* daughters."

Sir Thomas Browne had four daughters and no sons.

Sir James Mackintosh—a distinguished lawyer—married when young and in great pecuniary straits. He was living in the family of Dr. Fraser, in London, where Miss Catherine Stuart, a young Scotch lady, was a frequent visitor. She was distinguished by a rich fund of good sense and a warm heart, rather than by her personal attractions. An affection sprang up between them, and they were privately married, greatly to the offence of all their friends. She died a few years later, leaving him *three daughters*. She was distinguished for no intellectual qualities, but for plain, domestic, wifely virtues.

Beethoven's father was a drunkard, and left a family of *boys*.

The poet Thomas Campbell was proverbially indolent. He was the father of *two sons*.

In a volume of Spark's *Lives of Distinguished*

Men, which contains sketches of Count Rumford, Z. M. Pike, and Samuel Gorton, I find the first, at twenty, married a rich, elderly widow, and had one child by her—a daughter, as might have been confidently predicted; the second had three girls and one boy; and the last named, six girls and three boys.

The Rev. P. Brontë married Miss Branwell, member of a consumptive family numbering four girls and one boy. He had by her one boy and five girls, Charlotte Brontë being one of them. The mother lingered and died of consumption. The case is a very strong illustration of my theory.

Mrs. Southworth, the novelist, was married to a man who proved to be cruel, worthless, and improvident, abandoning her, after two years, with a *son* to care for.

Caroline Norton, granddaughter of R. B. Sheridan, was one of the three talented sisters—"the three graces." The *New York World* says of her: "The girl-wife—only nineteen—inherited her grand-parent's talent, and had already given proof of striking poetical genius. She married the Hon. George Norton, who was her equal in birth only. They had *two sons*."

Mrs. Felicia D. Hemans, at the age of nineteen, married Capt. Hemans, who was probably her senior by several years, and not in the best of health. They lived unhappily and were separated after some years. Little that is favourable seems to be said of him. They had *five sons* and no daughters. Mrs. Hemans, judging by her portrait, was a woman of the very highest order. She is described as "the most touching and accomplished writer of occasional verses that our literature has yet to boast of." The fact that

her husband was of infirm health is fatal to the
" physical vigour " theory.

Oliver Cromwell had *two sons* and *four daughters*.
He possessed an ample share of the lymphatic temperament, and yet other qualities were so much in
the ascendant as to make him a great man, mentally
as well as physically.

Junius Brutus Booth had *four sons* and *two*
daughters. He is described as " a real genius—
acknowledged to be such by all—of noble Roman
classic features, the highest type in short—a Michael
Angelo : yet he became a slave to the cup." I have
no reason to question this account, as it harmonizes
so well with the sex of his children. Such a man,
with good habits, marrying a lady of average ability,
would naturally have daughters ; and a husband of
no remarkable capacity with Booth's habits would
have sons, even if united to a wife of somewhat
inferior talent.

Thomas Jefferson had *six daughters*. The mother
of William Lloyd Garrison was a remarkable woman ;
she had a drunken husband, and a family of *boys*.

To the mind of any practical temperamentalist,
of whatever school, the strongest proof of the truth
of my theory would be that my standard of superiority perfectly accords with his ; judging, as he
does, by temperament ; my estimate is in perfect
harmony with every classification I have ever seen.

The case of Horace Greeley, with his *three sons*
and *four daughters*, was a perplexing one until I
understood the question of temperaments. Without
any knowledge of his wife, I looked for nothing but
girls from one of his great intellect. But his

temperament was of the encephalic, or cerebral, and sanguine,—a combination not so remarkable for power as for delicacy.

The case of Abraham Lincoln was a perplexing one at first. He had *four sons*, and this may be accounted for as follows. His wife was ten years the younger, which told in favour of sons. Previous to his active political life, which called forth all his latent energy, he was not noted for his mental activity. Indeed, at any period it may be truly said that his "*moderation* was known unto all men." His own words were, "I step *slowly*, but never step backwards." Mr. Frederick Douglas, in describing an interview with the President, said, "I was politely ushered into his reception room, and as he saw me approaching he began to draw up his feet preparatory to rising ; ere long he commenced to rise, and as I advanced he continued to rise," etc. This affords a striking contrast to the movements of an active nervous man, who would have risen on to his feet the instant he made up his mind to do so. During the critical times of his presidency, the responsibility of government also called forth his latent powers, and developed energy to an incredible degree. Had such events occurred twenty years earlier in his life, his family would, I doubt not, have comprised some daughters.

The charming little group of "Longfellow's Daughters" is familiar to every one. A man of the deceased poet's cultured mind and refined, active mental habits, having *one son* and *three daughters*, is an excellent illustration.

A recent case in St. Louis is interesting in relation to races. Ah Lee, a Chinaman, married an Irish girl. They have *two boys* and *two girls*, and so far the assimilation may be considered perfect. But the sons are as unmistakably Irish as the daughters are little Celestials. Assimilation was far from complete, and each sex stamped its opposite.

It is generally known that there is a notable difference in the cerebral development of various races and nationalities. Professor Morton, out of a collection of more than six hundred skulls, found, as may be seen by his *Catalogue of Skulls*, that the average size of German, English, and Anglo-American crania is ninety-two cubic inches; the average of ancient Egyptian skulls, eighty inches; of American Indian, eighty-four; and of Negro, seventy-eight.

In a majority of instances, were Americans to take wives from among the English, their offspring would be daughters, owing to the greater activity of the American temperament. Among the Germans, the result would be nearly the same, for a similar reason. A much smaller excess of females would probably result in uniting with the vivacious French. Were any of these nationalities to seek wives among the Spanish, Italians, or Portuguese, an excess of daughters would almost certainly be the result, as a vast majority of the women of Spain, Italy, and Portugal receive but little education, and are averse to mental effort in any way.

No official statistics are available for these comparisons of races, and their relative " superiority " cannot yet be determined. Still isolated cases are interesting, and point to the probable character of the result could we extend our researches. The instances I have given have deeply interested me, since they forcibly prove that no one condition or quality determines the sex, but that Nature takes account of all, and awards accordingly. As time, distance, weight, power, velocity, etc., in mechanics are interchangeable, insomuch that we can accurately calculate and substitute for the loss of one its equivalent of another, so is it with the vital forces.

As conclusive evidence, to my own mind, of the truth of my theory, I cannot do better than close this chapter with a somewhat more extended examination of the family history of the immortal Milton, already mentioned as one of the best possible typical cases.

He was a man pre-eminent in all the walks of life,—a poet, a statesman, a political writer, a musician, and a scholar of the highest order. There can be no doubt as to his genius, and this, with his active and temperate life, makes it interesting to trace his genealogy and his posterity. His gifted mother married an eminent musical composer, by whom she had *two* sons and *one* daughter. As might be anticipated from what I have said concerning the musical temperament, of these three children, Christopher had *one son* and *two daughters*, John had *three daughters* only. His sister married a public official, by whom she had *two sons;* he died, and she married his successor, by whom she had *two daughters*. Whether difference in age or enfeebled health on her part caused the other sex to result from the second marriage, it is impossible now to determine. John Milton's grandchildren consisted of *seven sons* and *three daughters*.

The sum total of the three generations will therefore be *two sons* and *one daughter* for the first, *three sons* and *seven daughters* for the second, and *seven sons* and *three daughters* for the third; making a total of *twelve sons* and *eleven daughters*—as nearly even as it is possible to be. Thus the law of balance continues the compensating process throughout succeeding generations of the human race.

But let us turn to the author of *Paradise Lost* himself. I quote from Johnson's " Lives of the Poets " :—

" At thirty, Milton was at Florence, where he found his way into the academies, and produced his compositions with such applause as appears to have exalted him in his own opinion, and confirmed him in the hope that, by labour and intense study, which, says he, ' I take to be my portion in this life,' joined with a strong propensity of nature, he might ' leave something so written to after times, as they should not willingly let die.' . . . What we know of Milton's character in domestic relations is, that he was severe and arbitrary. . . . ' Paradise Lost ' is a poem which, considered with respect to design, may claim the first place, and with respect to performance, the second, among the productions of the human mind. . . . The thoughts which are occasionally called forth in its progress are such as could only be produced by an imagination in the highest degree fervid and active, to which materials were supplied by incessant study and unlimited curiosity. The heat of Milton's mind may be said to sublimate his learning, to throw off into his work the spirit of science, unmingled with its grosser parts. . . .

" He had now been blind for some years; but his vigour of intellect was such that he was not disabled to discharge his office of Latin secretary or continue his controversies. His mind was too eager to be diverted, and too strong to be subdued. . . . The highest praise of genius is *original invention*. . . . Milton was naturally a thinker for himself, confident of his own abilities, and disdainful of help or hindrance. . . . His great works were performed under discountenance, and in blindness; but difficulties vanish at his touch : he was born for whatever is arduous, and his work is not the greatest of heroic poems, only because it is not the first."

I look in vain for Milton's superior in the scale of manhood.

This case seems to me admirably adapted for the summary of the whole subject as I would state it. Here we have the determination of sex in this single family, strictly according to the principle I have laid down ; and on the other hand we see, in spite of the originally partial inequality between the progenitors —the mother being "superior"—that the principle of self-adjustment in the course of two generations all but balanced the sexes once more; and this, be it remembered, is based on a very small number. If this be true of one family, it must be also true, with greater certainty and accuracy, of a whole nation, because averages become more exact ; and it is to the perfection of this principle of self-adjustment that I confidently appeal as the strongest verification of the theory laid down in this volume.

CHAPTER X.

PRACTICAL RESULTS—CONTROL OF SEX.

The subject practically considered—The useful results of its application—
Its influence on daily life—Its application to the unmarried—The
power of ensuring sex of offspring—Special directions to be followed—
Adaptability of the theory to every class—Physiological equality
desirable—Importance of avoiding great inequalities—Instances of
unequal distribution in families—Disadvantages arising therefrom—
Ideal home-life—Reflex influence of equality of the sexes in the family
—" Superiority " caused by external influences—Slight causes affect
sex of offspring—Sex definitely ensured—Condition and habit all
powerful—Easily altered, as instanced by athletes—"Superiority" not
to be lowered, but " inferiority " raised—The future child's interest at
stake—How its welfare may be attained—How to treat excess in the
nutritive system—Influence of exercise—Change of scene—Occupation
—Rest and recreation—Excessive intellectual work to be given up—
Temporary " superiority "—Hints as to obtaining it—The ideal wife
—Her home influence—Influence on her children—Useful lessons
drawn and illustrated by forty portraits—" Superiority " how attained
—Bearing of physical health on " superiority "—"Superiority " practically within parental control.

AFTER having concluded in the last chapter our
consideration of the theory of sex in its scientific
bearings, and having established its truth so far as
the evidence yet obtained will admit, by showing its
frequent and striking confirmation when confronted
with all classes of facts, I will now take up the more
practical side of the subject, and show the tangible
and useful results which will accrue from a direct
application of the theory to the circumstances of
daily life.

I have said that it will enable any one with a very
little practice to predict with tolerable certainty the

sex of the offspring of any given marriage; we have now to consider how it will aid us in going one step further, and that a most important one, *i.e.*, in the actual control of sex.

To those who are still unmarried the advice conveyed by my theory is simple enough, and the knowledge gained by perusal of this work is extremely easy of application. If a man desire that in his offspring there shall be a fair proportion of sons, he has only to make sure, in selecting his future wife, that she exhibit some of those traits, for the most part easily recognizable, which I have shown to be indicative of the requisite physiological "superiority," so as to insure that the difference may be within personal control. If he is himself of an extremely active temperament bodily and mentally, he should seek a wife in whom similar traits are present, and besides this it will probably be well for him to follow some of the directions on this head subsequently given in this chapter; and the same advice will need but slight modification to render it suitable to other less definite cases.

But before proceeding to treat of these, I would point to the desirability of choice of a wife who shall somewhat approach to physiological equality with the husband, and *vice versâ*. For apart from the fact that extreme physiological inequality is likely to be associated with other incompatibilities, and that it will be far better for the happiness of the parties themselves if there is no very great inequality between them, there is the equally important consideration of the future family life.

In my lists of several hundred households I find a

balance of the sexes in about one-third of the families, while another third have a large excess of boys, and the remaining third nearly as great an excess of girls. This gives no doubt on the average a near approach to a numerical equality; but the distribution should not be thus unequal, for it is not so in a state of nature, nor among animals. And among human beings this unequal distribution is productive of many disadvantages, for a family in which all the children are of one sex does not afford the proper ideal of home life. We should not look for the best example of domestic happiness either in a club-room or in a ladies' school. If the family be composed wholly of boys, they are apt to grow up coarse and unmannerly, and destitute of the refinement of feeling as well as of manner which results from feminine society; if it consist exclusively of girls they will become shy and prudish, narrow and petty in their aims and sympathies. All one-sided developments should be strenuously avoided if we wish for off-spring physically and mentally vigorous.

How then are we to ensure the desired equal distribution of sexes in our families? I reply, by following nature's method of procedure, and varying the relative " superiority " of the parents. Nature does this continually, as is plainly shown by the existence of families in which both sexes are represented. We have only to imitate nature so far as we can do so, and the like results will ensue.

The question then arises, How far can this theory be made to apply where the parents are already married? in other words, how far is it possible for parents to decide the sex of their future offspring?

To those questions I reply, briefly but confidently, that in the vast majority of cases it is quite possible. How this is to be accomplished I shall now proceed to show.

But the fact just mentioned, that in about one-third of the families tabulated an equal proportion of both sexes was found, may seem to militate against my whole theory. The reader may ask how do I reconcile my idea of "superiority," which I have shown to be temperamental and so far therefore permanent, with this evident alternation of the sexes, and the concurrently necessary alternation or change in the determining cause?

At first sight this no doubt appears strange, but like many other difficulties it will be found on examination to be really a confirmation of the theory it seems to oppose. Where the parents are in temperament and character nearly equal, very slight causes will suffice to turn the sexual scale at the critical moment, and hence one will be likely to be "superior" for the time being to the other, both being affected by the same causes of change, whether subjectively from change in themselves or their state of health, or objectively from without in the form of fatigue or other lowering causes. Those who have followed me through the preceding chapters will have no difficulty in understanding how slight causes may suffice to determine the balance, and I shall return to this subject a little further on, and examine some of these causes more in detail.

But to take up once more the question of sex in individual families. In the first place it can hardly be denied that what has been produced once can certainly be produced again under the same essential conditions, so that all parents who have had a healthy child of the sex they may now desire, can easily secure the same again if they but conform to the required conditions, and bring themselves to the desired position in relation to each other which they have previously held.

In the above mentioned list of several hundred families I find only twenty-four instances, or 4 per cent., in which the offspring were all of one sex. Hence it seems beyond all question that in the great majority of cases the sex of the children was at the option of the parents, had they but possessed the requisite information ; and it is to be remembered in this connection that nature is so far on our side in the endeavour to procure an equality of the sexes in a family, that this is substantially her rule, when unhampered by artificial social influences. Pigeons mate, and almost invariably produce two young, a male and a female, and I have occasionally found nests or litters of undomesticated animals, and in such cases have always been struck with the balance of the sexes shown. But how can any change be made, it will be asked, in the relative condition of the parents? The " superiority " spoken of is said to reside in temperament, and in the existence and activity of a certain mysterious life-force of which we know only the manifestations or effects. In short, it is an affair of constitution, and what can we do to alter that ?

Much, I reply—quite sufficient, at all events, to attain the desired end ; for although *constitution* makes the essential difference, and primarily determines relative "superiority," yet *condition* and *habit* are two very great factors in determining the actual result,—great enough, when properly used, to override the difference of constitution when this is not too pronounced. And it is by acting upon these that we are enabled so far to influence that result as to bring about a fulfilment of our wishes.

To make it clear that we can materially alter ourselves by the adoption of special means, I need only point to the effects of habit in acquiring dexterity in particular trades, so that what appears an almost miraculous performance to the spectator is a merely ordinary exertion to the trained workman; and again to the marvellous results achieved in a short time by training as it is practised by athletes with a view to raising their superiority in the matter of physical strength and endurance. The immense increase in power gained by a few weeks' preparation is familiar to all; while we know how fearfully potent is illness or alcohol to reduce one's condition so as to quite override temperamental character for the time being; but, as already observed, this latter must not be used as a factor in determining sex. Now the relative physiological "superiority" can assuredly be altered by analogous processes—within certain limits; so that just as the parents of girls and boys have their relative "superiority" changed from time to time by temporary occurrences, such as exaltation or subjection of health, pressure of work, lapse of time, etc., so we may produce a similar alteration by artificial means.

When sons only are born into a family, the cause may be traced to what I have termed "superiority" on the part of the mother, or "inferiority" on the part

of the father relatively to one another, the " superior "
parent stamping the opposite sex, as already explained
in the previous chapters. How then shall this relation be reversed in order to ensure a daughter? As
a rule I should say it is not desirable that the
" superior " parent should be lowered to the level of
the " inferior," but rather that the " inferior " parent
should have recourse to the elevating and exalting
method in order to secure the desired sex by acquiring
a temporary " superiority." For although it may be
thought easier to lower than to raise oneself in the
scale of energies, yet for the sake of the future child
surely the more exalting process should be resorted
to, even if it require a somewhat greater effort. Few,
I venture to predict, would be found who would not
adopt the method of raising themselves in the scale
in order to ensure the inheritance by their offspring of
the best qualities it is in their power to transmit.
Stock-raisers say, " We always breed *up*, not *down*."
Will parents be less dutiful to posterity?

We may then look a little more closely into the
methods to be employed to raise an " inferior" parent
to a higher point in the sexual scale, so as, for a
time, to make him or her the actual "superior," and
thus to determine the opposite sex. I have said that
temperament is a permanent factor, only alterable in
a very slight degree, and that therefore we must look
to change of *habit* and *condition* to bring about the
desired result. If the nutritive system be absorbing
an excess of vitality, it will be clearly necessary to
incite the nervous system by gentle physical stimuli,
such as cold or shower baths, athletic exercises, and
by any other means calculated to exalt its condition.

This greater mental activity will bring about a simultaneous physical readjustment, and by diverting the excessive blood-supply from the nutritive section, and directing it in a fairer complement to the other parts of the system, will raise the nervous and intellectual system into a fuller prominence. If the lymphatic system be excessive, exercise must be taken, the indolent inclination must be conquered, and an appetite acquired, so that the nutritive system may receive a stimulus, the pulse be made to beat more actively, and the nervous system be vitally quickened. If the sanguine temperament have exhausted the system by excitement, then quiet mental exertion, such as contemplative reading, or routine brain work, will give the nervous system the necessary time to reassert its ascendancy, and a more composed habit of life will give a healthier tone to the nutritive system.

From these general outlines of what is necessary an idea may be gained of the purport of the following suggestions as to the detailed means to be employed; and the directions and the reason for the various specific hints will be readily understood by all who have carefully perused the earlier chapters.

Let us first take an example most likely to occur in practice—that of a father who is desirous of having a son. We will suppose he has one or more daughters, thus showing that the only impediment is his " superiority," in our special sense as affecting sex, to his wife. We may reasonably suppose that he is of the nervous type, an active, energetic man with a strong will, well-marked features, and possessed of other characteristic marks of " superiority," as given in Chapter VII.

Such a man may without fear of detriment to his off-spring safely reduce his superabundant energy in some measure, and deflect vital attention towards his nutritive system. Indeed such a man very frequently has his cerebral preponderance *morbidly* raised by over-stimulation of the brain, so that it is rendered abnormally active by the high pitch of nervous tension to which his whole system is strung under the pressure of intellectual work, and the other systems of the body are consequently relatively injured and reduced. He will frequently have a weak digestion if not poor health. Now here it will be a real gain to the constitution of his child if he subdues himself for the time being, so far as cerebral activity is concerned, and at the same time recuperates his health and digestive powers by change and rural recreation, forgetting meanwhile his mental duties. He should leave these and indulge in physical exercise, such as moderate riding, easy hunting, yachting, etc.; and by these means and a large measure of fresh air acquire a healthy appetite. He will thus bring himself to a higher animal condition, and yet not lower his intellectual status beyond the normal standard; and he will thus be much more likely to beget a son, while, at any rate, he will not transmit to his offspring the heritage of a morbidly active brain, which is usually attended with many penalties to its possessors, and not unfrequently leads to insanity.

He should rest his brain for a month or so, whilst cultivating his health, physical strength, and vigour. Let him seek to build up his frame by judicious out-door exercise, prolonged daily as far as may be without actual fatigue. In addition to riding, etc., already mentioned, I may suggest rowing, riding a tricycle, pedestrian tours, mountaineering, etc., according to his strength and inclination. He must sleep as much as possible, and avoid exciting subjects, whilst living well and temperately.

All this will lessen his abnormal " superiority " without lowering him in the scale of real stamina, as would be the case if his " superiority " were reduced by illness, drink, or other excess, or any depleting or enervating cause. Anything tending to enervate should of course be avoided, as none would wish to have a son by lessening their own vigour in a manner that would tell adversely upon such offspring.

The husband in such a case may then safely subordinate himself so far; but the wife must aid in the matter, and must raise her standard by every means in her power. The exact means to be employed must largely depend upon the special circumstances and temperament of the individual. But in civilized societies the life of woman tends to develop the nutritive and lymphatic systems, and in most cases the remedy will be to stimulate the nervous energies, and overcome the sluggishness due to predominance of lymph. A simpler diet,—and more sparing in some cases,—cold baths, exercise, and active employment should be resorted to, till the weight, if at all excessive, be somewhat reduced, the circulation strengthened, and the brain thereby reinforced. Some study or attentive reading may be taken up with advantage, and all means employed to awaken and energize the mind and subordinate the nutritive system, or rather to utilize the food-supply for higher purposes. All strong and frequent tea and coffee should be abjured when thus seeking to gain a "superiority" for the purpose in view.

In the converse case, where sons predominate and a daughter is desired, the husband must strive to reduce any tendency to corpulency, dispense with rich dishes, cigars, and wine, strong tea or coffee, eat and sleep less, drive less and walk more, undertake some real work, and fight against the fatigue he will assuredly feel at first, and which shows that the nervous resources are exhausted by the demands of the nutritive system. Let him, in short, do all in his power to stimulate his sluggish nature, and encourage abstemiousness; and he will probably be rewarded by the birth of a daughter.

In suggestions to the wife, it seems hardly necessary to repeat what has already been said about the means for reducing abnormal nervous energy; but I may still give a few additional hints as to the way in which she should act in order to obtain a daughter. We will suppose that she is rather nervously active, and thus probably of slender figure. To her I would say, sleep as late and as much as you can, take things more easily than you have hitherto done, cease all mental effort till you begin to gain a little in weight every week. No matter if you do not read all the magazines for a month or two, nor finish the usual number of books per week from the circulating library, nor answer your correspondents with wonted promptness; you have an object to gain, and can afford to rest Do not fear that this apathy or indifference will be transmitted to the hoped-for daughter, for you will have nearly all the period of gestation in which to exert every faculty, thus moulding the plastic germ of life at your pleasure.

Since a full possession of nerve-force is essential to " superiority " in the casting of sex it may be well to mention here that modern research shows us that this nervous energy is largely developed during the assimilation of oily and amylaceous foods, that is, of digestible fats and starches, etc., and which are more especially heat-producers, as distinguished from the nitrogenous or flesh-forming foods, such as the gluten of flour, lean meat, eggs, etc., though as these latter also contain some fats and produce a certain amount of heat in their combustion, they must not be neglected on this score. Also let phosphates be partaken of more freely than usual with or in the food,—these phosphates being comprised in the best fish, in oysters, in lightly-cooked eggs, in lean, underdone, tender beef, in decorticated bread, made of entire wheaten flour, and being further procurable as concentrated phosphates from the manufacturing dietetic chemist. In short, any one suffering from nervous exhaustion should, in order to attain " superiority," adopt for two or three months a diet in which fat and oils, butter, cream, milk, cocoa, etc., form a considerable item—so long as they do not disagree.

There will be a very large number of abnormal cases which require special diet, and indeed every individual case should strictly be judged separately; but I may here call attention to a method of treatment introduced by S. W. Mitchell, M.D., of Philadelphia, advocated especially for certain refractory cases of nervous exhaustion and anæmia

with consequent loss of appetite and energy, constant
fatigue, and hysterical symptoms. These cases, not
unfrequently found among women, are more easily
recognised than described, but the patients, though
they may not be actually ill, will infallibly be physiologically " inferior " to their husbands, and a special
course of preparation will be needful for such to
ensure sons. The treatment to which I allude is
described in a little work by Dr. Mitchell,* and consists of complete rest in bed for a period of several
weeks, no exertion, mental or physical, being allowed,
but an artificial excitation of circulation being
obtained by rubbing, pinching the skin and surface
muscles, flexing and extending the limbs, etc.,
whereby a local stimulus is given to the small nerve
fibres and the various networks of capillary blood-
vessels in every part, and functional activity is restored
or increased without any physical fatigue or nervous
strain. At the same time a simple diet—such as
milk—exclusively for the first few days, restores tone
to the digestive functions, and other food is soon
desired, which may be judiciously added at the
patient's pleasure, until the appetite and digestive
capabilities are at least raised to the normal state.
When this has been done nature will complete the
renovation; and from the absence of any drain upon
the reserves of vital force, while the food supply is
increased, and far more thoroughly assimilated, a
steady gain of flesh and blood will result, which is
an indication of a general constitutional improvement such as seems to me well calculated to
produce parental "superiority." This treatment,

* " Fat and Blood, and How to Make Them."

though somewhat novel, has been found, it is claimed, eminently successful in medical practice, and there can be no doubt that the results thus obtained run in the direction of our object, namely, the determination of sex, by conducing to " superiority."

On the side of the recuperating progenitor many special forms of treatment, such as the application of electricity, medicated baths, or sagaciously chosen tonics, may prove highly advantageous to individual constitutions; but here, from the very nature of the case, no general rule can be laid down; the circumstances of each inquirer must determine the course to be pursued. But the idea I have sketched being once fully grasped, there will be little difficulty in adapting the means to the end to be gained, while in most cases the directions given in this chapter will suffice for such purpose to any intelligent reader.

It cannot be too strongly emphasized that the treatment or regimen shown in this chapter to be needful for the securing of sex at will, tends at the same time in every case to improve the state of health and constitution of the parent who puts it into effect, and thereby ensures the best possible constitution for the child of the desired sex, thus securing a double benefit, while in no case is it possible that the adoption of either of the systems of physical training herein described can be productive of any injurious results. In nearly all cases, indeed, those who make the trial will experience a most decided improvement in their own health, and will be the gainers directly in themselves, as well as indirectly by securing the object of their wishes.

In cases where the inequality between the parents

is much less pronounced, I am strongly inclined to the belief that there is a much quicker method by which temporary "superiority" may be attained. If "inferiority" in determining sex would result from a sudden illness, or the reaction after a state of intoxication, why should not virtual "superiority," though in a less degree, be produced in a similarly short period? Take, for instance, a husband who is but slightly "inferior" to his wife, and who is desirous of a daughter. If a Turkish bath is known to have an exhilarating effect upon him, making him feel like a new man for the time, let him avail himself of such bath, the influence of which will, I believe, tend to produce temporary "superiority." Other methods will suggest themselves to the physiologist, and there are doubtless medicines or other agents that will further aid the desired result. Turkish baths are not everywhere available, nor always invigorating, and in such cases other baths may be found more serviceable to the same end. It is not unlikely that an opposite effect may also be produced on the other party, without injury to offspring, by some " Sitz " or other sedative bath.

Additional methods may be suggested by the medical adviser who has thoroughly mastered this subject, and who, as a physiologist, will more thoroughly understand what is best calculated to produce temporary "superiority" in the individual consulting him. The matter is made comparatively simple by the fact that the desired effect is only needed for a time, and that whatever reaction, if any, is experienced afterwards will not change the result. I may add a caution respecting the use of alcoholic liquors as a stimulant: alcohol stupefies the brain, and even in its first stage its use would probably produce the opposite result, and also lead to degenerate offspring, as alluded to in a previous chapter.

Whatever means are available on the part of the father to produce a daughter, I need hardly add, have only to be adopted by the mother to produce a son.

We shall be assisted further in the solution of the problem as to how the wife may elevate herself, by the following quotation from a small volume entitled *Bits of Home Talk*, from the pen of Helen Hunt, which gives an admirable delineation of the ideal wife :—

" It is everybody's fault that the average home is stupid, dreary, insufferable—a place from which the fathers fly to clubs, the boys to the streets. But when we ask who can do most to remedy this,—in whose hands it most lies to fight against the tendencies to monotonous stupidity and instability which are inherent in human nature,—then the answer is clear and loud. It is the work of women ; this is the true mission of women, their ' right ' divine and unquestionable, and including most emphatically the ' right to labour.'

" To *create* and sustain the atmosphere of a home, —it is easily said in a very few words ; but how many women have done it ? How many women can say to themselves or others that this is their aim ? To keep house well, women often say they desire. But keeping house well is another affair,—I had almost said, it has nothing to do with *creating a home*. All *creators* are single-minded. Never will the painter, sculptor, writer, lose sight of his art. Even in the intervals of rest and diversion which are necessary to his health and growth, everything he sees ministers to his passion. Consciously or unconsciously, he makes each shape, colour, incident, his own; sooner or later it will enter his work. So it must be with woman who will create a home. There is an evil fashion of speech which says it is a narrowing and a narrow life that a woman leads who cares only, works only, for her husband and children; that a higher, more imperative thing is, that she herself be developed to her utmost. Even so clear and strong a writer as Frances Cobbe, in her otherwise admirable essay on the *Final Cause of Woman*, falls into this shallowness of words, and speaks of women, who live solely for their families, as ' adjectives.' In the family relation, so many women become even less, that human conception may perhaps be forgiven for losing sight of the truth, the ideal. Yet in woman it is hard to forgive it. Thinking clearly, she should see that a *creator* can never be an ' adjective '; and that a woman who creates and sustains a home, and under whose hands children grow up to be strong and pure men and women, is a creator. Before she can do this, she must have a development; in and by the doing of this comes constant development; the higher her development the more perfect her work; the instant her own development is arrested, her creative power stops. All science, all art, all religion, all experience of life will help her. Could she attain the utmost of knowledge, could she have all possible human genius, it would be none too much. Reverence holds its breath and goes softly, perceiving what it is in this woman's power to do;

with what patience, steadfastness, and inspiration
she must work.

"Into the home that she will create, mono-
tony, stupidity, antagonisms cannot come. Her
foresight will provide occupations and amusements;
her loving and alert diplomacy will dispel disputes.
Unconsciously, every member of her family will be
as clay in her hands. More anxiously than any
statesman will she meditate on the wisdom of each
measure, the bearing of each word. The least
possible governing which is compatible with order
will be her first principle; her second, the greatest
possible influence which is compatible with individuality. Will the woman whose brain and heart are
working these problems as applied to a household be
an 'adjective,' be idle? She will be no more an
'adjective' than the sun is an 'adjective' in the
solar system; no more idle than Nature is idle. She
will be perplexed; she will be weary; she will be
disheartened sometimes; but she will never withdraw
her hand for one instant. Delays and failures will
only set her to casting about for new instrumental-
ities. She will press all things into her service.
She will master sciences that her son's evenings
need not be dull. She will be worldly-wise, that her
husband and daughters may have her by their side
in all their pleasures. She will invent, she will
surprise, she will forestall, she will remember, she
will laugh, she will listen, and she will be three times
loving."

As to whether the nutritive or intellectual system
should receive the more attention in the case of
women, there is a difference of opinion; but there

can be no doubt that the nutritive system must be thoroughly developed in youth, and properly cared for throughout life, as the health and constitution, and hence the happiness, of offspring depend upon the soundness of the mother's nutritive system. There can be no objection to increased intellectual culture, provided this primary object be kept in view; but there seems to be some danger of our losing sight of it in the present eager advocacy of the higher education of women,—indeed, I may cite Mr. Charles Roberts, M.R.C.S., who believes that we may already trace the ill-effects of over development of the cerebral system in women in the increase of male still births, as he states in the *Lancet* of December 11th, 1880.

But some faint-hearted ones may say, "The vital predominates in me; I can never become the superior of my consort and obtain the sex I desire." I maintain the contrary. The following advice to a patient as to how to get well is by Dr. J. C. Jackson, of Danville, New York, a practical medical man, and shows what may be done towards gaining the coveted end by rousing one's energies :—

"Now you can be well if you choose. To do so, however, you will have to change your conditions of living largely, and be thoroughly revolutionized. When a man has such constitutional predispositions as you have, there is no help for him in little detached efforts, nor in any sort of patchwork processes. You are to go to the bottom and touch the very primal conditions of your life. You must affect, under your conditions of living, those organs which have for their special office the turning of food into blood, and so the building up of tissue and the keeping of such tissue built up healthfully. If you do not, then all your efforts, being but superficial, will only produce superficial results, and you must, under such circumstances, have poor health, gradually increasing feebleness, and perhaps thorough and ruinous decay at no distant period. I would make myself over again if I were you. It can be done. To live differently from what one has been living, is to be different from what one has been; and this then can have as direct reference to bodily organization as to character. I know that in this respect I shall be considered as imaginative. Nevertheless, I know that what I say is true,—that it lies within the power of the individual to modify his physical organization so that, in respect to the exhibition of qualities or powers, it may be very different from what it has been. Weak men may become strong; thin men may become fleshy and bulky; the bulky and fatty man may become thin and spare; the scrofulous man may rid himself of his scrofula; the consumptive man, if his organization remains still entire—no structural lesions having taken place—can thoroughly rid himself of his consumption. "What can thus happen in physical conditions can, to an equal extent, happen in mental conditions. The dull man can become bright, the excitable man cool; the impulsive man reflective; the thoughtless man thoughtful; the careless man heedful; the inconsiderate man careful; the cross man pleasant; the hateful man genial and neighbourly. Man has in him as an ever-present and plenary power, the force, under right conditions, to bring about all these changes. If, therefore, you wish to get well, do not sit down with your dizzy head and your dull brain and your defective circulation as though you were helpless, but help yourself." This amount of resolution, despondent reader, would be far more than is requisite for producing sex at will.

In the case of the lower animals, to obtain our choice of sex we give attention chiefly to age, exercise, feeding, the time of year, etc., because we have no intellectual or moral qualities to work upon; this enumeration of the manifestations of life presents the only accessible avenues we have to the fund of vitality existing in the brute creation. But with human beings the case is very different; man is capable of amelioration almost indefinitely. Knowledge is power in every sense. A parent's knowledge is not transmitted to his or her offspring, but *mental capacity* is bequeathed to each succeeding generation. The exercise of intellect enlarges the cerebrum, and all mental acquisitions necessitate a larger and continual supply of nerve force, and nerve force is, as we have seen, the principal element in deciding sex; for this reason intellectual development is conducive to individual " superiority."

With the hope of making plainer the *modus operandi* of obtaining sex at will, I shall now give a few hints with reference to some of the couples pourtrayed in this work.

In order for A 8 and B 8 to have sons—the status of the husband being estimated at 110, and the wife at 95—I need make no suggestion to him, provided his habits are good, but for her there is plenty to do. She is a departure from the ideal of woman, and should try to remodel herself: her abnormal tendencies have become a second nature to her. Her weight is evidently excessive, but she is not a large eater, for, owing to her lymphatic tendency, there is no great demand upon the vital resources. Debay would recommend highly-seasoned food and good wine to stimulate her sluggish nature, but this course would only further stimulate the *nutritive* system, whereas the *nervous* is the one to be developed, if we wish to raise her above her husband's standard of "superiority." Inactivity is one of her great defects, and she must therefore secure active mental and physical exercise. It will require some effort to make a commencement, but it is only the first step that is difficult. She should take cool or cold baths to quicken the general circulation and invigorate the system; these will also serve as a species of, or substitute for, exercise. She must endeavour to materially reduce her excessive weight, and eradicate as far as possible the tendency to excessive secretion of lymph; to assist Nature in that direction, and to strengthen the nervous system, she would find it advantageous to be very abstemious for a week or two. She should become enthusiastic over some intellectual occupation: if music be her forte, let her seek to excel as a composer, and thus stimulate her mental energies. The best physical exercise may be for her to walk a number of miles daily— even a dozen miles would not be excessive after practice for a week or two. I would advise her to become methodical in everything, and, in view of the amount of work to be done, never to yield to listlessness or languor. By faithfully complying with these suggestions for a few months, the desired ascendancy would doubtless be acquired.

I will now make some suggestions to A and B 16, to enable them to secure a daughter, the husband being perhaps 30 in arrear. To begin with, he has about fifty pounds of superfluous flesh, which should disappear in order to give tone to the nervous element. The vital energies are capable of producing a certain amount of force daily, but if so much of it must be spent in supporting the physical frame, but little progress can be made in developing the cerebral system. Either a finer quality or a greater amount of vitality is required for some operations than for others, as may be easily demonstrated. For instance, the process of digestion consumes nervous force ; so do running, bathing, and thinking, insomuch that, if one of these be persisted in after eating, indigestion is sure to ensue. On the other hand, conversation, singing, walking, or almost any one of the physico-mental operations, serves as a grateful stimulus to the digestive organs. With most people, social conversation aids digestion, while argumentative discussion arrests the process altogether. This fact I consider as demonstrating that the intellectual powers are consumers of a greater quantity or a finer quality of vital force than that which we require for the social faculties. In confirmation of this view we find that persons with large social powers have a coarser organization than those with predominant moral and intellectual faculties, and *vice versâ*. I dwell upon this point to show A 16 the necessity of stimulating his cerebral energies in order to gain his object.

Besides the vital force necessarily expended in keeping every human frame in good order, there is generated a surplus fund of life which we can expend in ways more or less profitable. If it be expended in manual labour, we obtain increased muscle, besides the work accomplished with our hands; if in feasting and debauchery, we consume every vital resource in unloading the system and repairing wasted tissues, having little else than corpulence to show for it all. But we may, if we choose, invest this surplus vital force in adding largely to our cerebral system and its activity.

Generally the muscular system is made to work in order to minister to the nutritive system, rather than to the nervous, but development of the cerebrum is certainly unequalled as a husbanding of life's surplus funds. The financial maxim that "what you spend you have, but what you save you lose," can never be applied to increase of brain power. The muscles and the abdominal viscera are possessed of either a small quantity, or a weak quality, of nerve force or electricity, as compared with the brain; and the lower portions of the cerebrum, also, as compared with the upper, or moral and intellectual parts, which seem pregnant with the very quintessence of life. The advantages to be secured by reducing the waste of our frame to a minimum, and of devoting the surplus to the elevation of our minds, is apparent. Therefore, if A 16 wishes to have daughters, I advise him to discontinue the use of tea, coffee, wine, tobacco, condiments, and all other stimulants which but minister to the appetite; he will then begin to lay up a store of vitality in the cerebrum. There is sufficient vital energy in his frame, and it only needs to be directed into the proper channels to elevate and ennoble him; to make him the " superior " of the one by his side, and the father of daughters.

For A 2 and B 17 to obtain daughters, much of the foregoing advice is equally applicable; the most striking difference in the two individuals is, perhaps, the moderation of the latter. He is, judging from his epaulets, a military man, and in that capacity he should know that many a battle is lost from an inability to rally and concentrate troops promptly. He has never drilled himself to summon all his nervous energies, to concentrate all his vital forces at a moment's notice. An anatomical examination of his frame and that of his companion would probably show a great excess of nervous force in the latter. His companion is not selfish or niggardly in anything she undertakes, but generous and whole-hearted, throwing her very soul into every enterprise. Let it be remembered that the basis of " superiority " is the *quantity* and *quality* of nervous force a person is able to command at any given time —and, generally speaking, the amount one is in the habit of commanding.

A and B 17 differ nearly 50 in the scale of superiority, and considering the age and temperament of A, I question his having the resolution necessary to obtain a daughter. Most of the faces coupled together in this work represent greater extremes than are usually met with in real life; and it would therefore be useless to enlarge upon their individual possibilities.

I trust that the foregoing will be **found explicit**

as it is essential here that there should be no misapprehension. It will be seen that the aim should be to strengthen those parts of the system which either from temperament, disuse, or long habit have become weakened or impaired, while at the same time checking morbid activity in any direction. I would not be understood to maintain that only the nervous system should be cultivated,—indeed any one-sided development is pernicious, and the lymphatic system may well be encouraged in those of highly nervous temperament to check the morbid restlessness which becomes an element of inferiority by permanently injuring the general health.

So each system should be developed to aid in producing that balance which we call health, while the nobler part of us, the nervous system, should of course be in healthy activity.

Our " superiority " will as a rule be greatest when we are at our best physically and mentally, and to ensure sex at will, the condition of the determining parent should be raised to the highest point of efficiency short of morbid activity, of which his or her system is capable.

It will of course be understood that general physical health and constitution is also an important factor in its bearing upon the result. A fine specimen of the sanguine temperament may be, in the individual case, the actual "superior" of another individual of weak constitution and enfeebled health, though of the strictly nervous temperament, and so of others.

Illness is perhaps the most effectual agent in lowering the " superiority " ; and children conceived after an illness will be assuredly of the sex of the invalid parent, and will generally be delicate. Perhaps the nervous temperament, from its extreme sensibility, is affected sooner and more profoundly by any disease or derangement than any of the others, and the sons of many a highly able and energetic father doubtless owe their existence to the temporary lowering of his standard by a sick headache, acute pain, a sudden shock, or the slow attack of disease. A family of healthy daughters and one delicate son generally tells a tale of such description.

Not physical or genital vigour, nor passional energy, nor greater age, nor higher nutrition, but all of these and much more, as so many manifestations of greater nervous or vital power, really determine "superiority," and thus cast the sex of offspring; and very much of this lies within our own control.

In concluding these special hints, I confidently believe that a time will speedily come, when with a careful study of the matter, and, in special cases, with help from competent scientific and medical advice, a complete system may be laid down whereby either sex can be healthily and readily begotten by any couple in average health, who are not widely mismated; and this with a near approach to certainty.

CHAPTER XI.

OTHER SOCIAL PROBLEMS—ELEVATION OF THE RACE—
CONCLUSION.

ALTHOUGH I have now considered the practical bearing of my discovery on the main subject in hand, and may thus be said to have brought my task to an end, there is a yet wider application of my theory, from which most important results may be deduced,—results capable of solving Social Problems quite as important as that of sex. In this brief chapter I can do no more than glance at some of these, with the double object of justifying the title of the present work, and of strengthening the foundations of the theory I have sought to explain, by showing its beneficent influence on other and remoter parts of the field of enquiry. A theory which, while largely conducive to the increase of domestic happiness, can be shown to be productive of increased happiness to nations, by removing social difficulties, and by encouraging the procreation of a race continually rising in nobility and capacity, surely claims a candid examination at the hands of those qualified to form a decisive opinion.

The previous chapter shows how an observance of the rules required by my theory will raise parents to their best possible condition ; and this will not only react upon their offspring, but will also promote a similar result in many other ways. For instance, I trust it may appeal to our manhood. A standing disgrace to nearly all civilized countries is that there is an excess of some 5 or 6 per cent. of male births, but, to mention something still more discreditable, in spite of this the excess is insufficient to fill the gaps caused by an excessive mortality which is chiefly owing to the bad constitutions transmitted by fathers whose life stream has been sapped by indulgence or excess of one kind or another. What Mirabeau said of himself is true of only too many among us : " My early years have already in a great measure disinherited the succeeding ones, and dissipated a great part of my vital powers." Thus again says the Italian Giusti : " I assure you I pay a heavy price for existence. It is true our lives are not at our own disposal. Nature pretends to give things gratis at the beginning, and then sends in her account."

How many with noble brows, flowing beards, and the other insignia of virility might make a similar confession !

> " Real glory
> Springs from the silent conquest of ourselves,
> And without that the conqueror is nought
> But the first slave."

With a morality among men as strict as it is among women, there would not only be no male excess of births, but the males born would not die in excess as they now do, and consequently the lamentable redundancy of women between twenty and fifty would be largely reduced, and with it the amount of unhappiness that is its inevitable concomitant. Wars, accidents, and pestilence would for a time leave some female excess, but it would be gradually diminished.

This also appeals to the ladies, the mothers of the land, for they have it in their power more than any class to remedy existing evils. Philanthropists in vain discuss the reason and remedy for the redundancy of women. Noble-minded Christian women can remove this social blight if they will; the present volume plainly indicates the cause and cure. It will require an earnest effort and one generation to eradicate it, but it can be accomplished. Looking to the most civilized lands, we meet problems which may well challenge the attention of the philanthropist and Christian statesman. In Great Britain and Ireland, the female population in 1881, even including the army and navy, exceeded the male by 738,668. In the six New England States, the excess in 1880 was 92,556. These results are partly due to emigration, war, shipwrecks, and the risks and calamities which pertain specially to the employments of men —causes which will not cease, but continue—but chiefly to preventible immorality among men.

If, as already stated, there is in our own day an excess of some 5 or 6 per cent. of male births, owing to the greater amount of immorality among men than among women, we should expect to find

a far greater excess of males in past times, even a
century or two back ; for there can be little doubt
that there has been an improvement in this respect,
as in many others. The drinking habits of the upper
classes have been modified considerably, and the same,
though to a less degree, may be said of other forms of
excess. The same result should be more strikingly
manifest if we looked back to times yet more remote,
when the lawless nobility and their numerous retainers
had no bounds set to their indulgences save the lack of
power to gratify them, and it would be a curious and
instructive subject of investigation to endeavour to
establish this by reference to old registers, family
records, taxation returns, etc., which should show a
very large number of families with sons in the
majority. But there is one indirect line of evidence
which goes to prove that such an excess of male
births must have taken place, as our theory would
lead us to expect. If in our comparatively orderly
and peaceful days the 5 per cent. excess of male
births is more than obliterated by disease, accidents,
and wars, we must reckon a very much larger percentage of deaths from these causes, especially war, in
ages when national wars and insurrections were of
almost constant occurrence, and when private war
between neighbouring nobles, raiding, border-lifting,
and other forms of petty warfare were universal and
continuous. Doubtless in those times women were
also less moral than they are now, but the difference in
their case would be far less than in that of men, owing
to the life they then led of confinement and isolation
in the feudal castles. The greater excess of male
births should therefore still be found. And it

becomes almost certain that such an excess must have existed to meet the terrible drain above spoken of, more especially as we do not read of any complaints of the redundancy of women.

Again, the most successful men of those times would be of a rude and robust type, for personal strength and stature, and the possession of brute courage, would be considered the most valuable qualities in a man, while those of a gentler and more studious disposition would seek refuge in the cloister. The successful men would thus have a free choice of wives, and their rank and power would enable them to select from among the most accomplished and beautiful. Among women, the most likely to seek refuge in the convent would be those who were unable to secure the shelter of a home. In this way also we may see how there would be a tendency towards an excess of sons, owing to the father's "inferiority" on the one hand, and the mother's "superiority" on the other. I am not now speaking of a rule, but of a tendency, which would probably be evident in some families, but unmistakably so in the aggregate. In the absence of direct evidence from records, I think the immense male mortality in the middle ages from wars and private feuds justifies the presumption that a larger percentage of male excess than is now shown by civilized countries must then have existed.

But even at present there remains a most lamentable excess of women, and a great mass of evil and unhappiness is its mournful result. "In such conditions," says Mr. Henry W. Sage, "for this vast mass of humanity, what necessity for restraint and limitations, for elevation and purity, for the positive control of moral and intellectual forces over those of the gross animal nature! And how can these results be attained?

"There is but one answer to this question. By that elevation of character, by the broadening and deepening of the whole nature which comes from early culture and education. . . . And will she be less woman with riper development of all her faculties? As wife, as mother, as sister, companion, and friend, will she be less true to faith and duty? Is man made dwarf or giant by increase of moral or intellectual power? We all know what culture in these directions does for him. It will do no less for woman; and when as free as man is to seek her own path in life, whenever led by necessity or duty, hope or ambition, when opportunity and aid for culture in any direction are hers, then may we expect to see woman enlarged, ennobled in every attribute, and our whole race through her receive impulsion to a higher level, to all things great and good."

The condition of the next generation, and the destiny of our race, are therefore, to a very great extent, in the hands of the men and women of the present day. Doubtless, the principle of adjustment already explained will ever tend to produce a numerical equality between the sexes at birth, but if masculine immorality be not checked, the constitutions of our posterity will be more and more weakened, until the proportion of male children dying in infancy will be very large indeed. The existing discrepancy of the sexes amongst adults will therefore be likely to increase, and thus progressively to add to the number of enforced celibate lives, and to take from the sum total of human happiness.

I have dwelt upon one practical result of the adoption of measures for securing sex at will, viz., the rectification of the existing inequality between the sexes; but there is a wider ulterior result, which I commend to the attention of sociologists and all well-wishers of our race.

For just in proportion as male morality is improved must female culture be extended, in order that the mothers of the generation to come may be, in the aggregate, on an equality with their husbands, and that there may be a proper distribution of the sexes. If man is to be elevated, woman must be elevated also, or a disparity equally harmful in its results may be looked for in the other direction. Thus an amicable rivalry, or emulation in the path of progress and improvement, will be set up between the sexes, and this will contribute to the advantage of the succeeding generation.

The application of my theory on a large scale would thus be eminently beneficial in promoting the elevation of humanity. For the widespread and pardonable vanity of our nature is such as to impel men to seek for that kind of immortality which is open to all, viz., the perpetuation of our name and race, and this leads to a natural desire to have sons. Enlisting this potent ally in my favour, it is evident that "superior" mothers must be sought for, if we are to ensure sons to inherit our names; and this of itself cannot fail, by the laws of inheritance, to impress a nobler type upon our offspring. If the selection of wives were universally based upon such a principle, an appreciable effect upon society at large would speedily result. In any individual case this effect may be small, because many other influences affect a man and determine his choice, but the addition of a new factor to a large number of cases will assuredly tell upon the aggregate result, though we may be unable to produce a single instance in which the new consideration has wholly determined the selection. The question of sex occupies a somewhat analogous position : when several forces are nearly in equilibrium a very small additional impulsion will determine motion in any given direction.

But there is one way particularly in which this consideration is likely to have marked results. Men of undoubted superiority will more especially experience the wish to transmit the memory of their names to their sons, and such men will be especially likely to be swayed by scientific considerations. Now it is evident that their selection must be very carefully made, and a very " superior " woman must be chosen, if they are to realize their wish. But whether they succeed or not, good must follow ; the best will seek to mate with the best, and a race of progressive superiority, of very great excellence, will be the result. In short, the effect of the adoption in practice of stirpiculture as advocated in this work will be to improve the human race by a process exactly similar to that which the stock-breeder terms " breeding up," and exactly the reverse of that which obtains to a very large extent among ourselves at present, and which under the mournful name of " the non-survival of the fittest " has been so bitterly deplored by the late Mr. Greg and other writers on social economy.

We may well imagine that were the theory here advocated generally accepted, a public opinion would soon be formed under the power of which it would be a social disgrace to a man to have only sons, while a stigma would attach to all families exclusively of one sex. This would act as a spur to "inferior" parents to arouse their sluggish natures and to become active and useful members of society, for there is no point on which we are more sensitive than this of family; and in the light of this new theory our sensibilities will become still more acute.

Incidentally then to the main question of sex, it will be seen that many suggestions have been thrown out for the solution of other social problems; but the chief points in this work, and perhaps the most important of all social problems just now (because they lie at the root of many others), are the actual inequality of the sexes and their possible equality if both sexes were equally virtuous.

Mr. Greg, in his "*Literary and Social Judgments,*" believes that the chief reasons for women being in excess are, "the luxury of the age and the profligacy of men" (which is far truer than he ever dreamed), and he very justly says:

"Few men—incalculably few—are truly celibate by nature or choice. There are few who would not purchase love, by the surrender or curtailment of nearly all other luxuries and fancies, if they could obtain it on no cheaper terms. In a word, few—comparatively very few—would not marry as soon as they could maintain a wife in anything like decency or comfort, if only through marriage they could satisfy their cravings and gratify their passions. If their sole choice lay between entire chastity,—a celibacy as strict and absolute as that of women—or obedience to the natural dictates of the senses, and the heart in the only legitimate mode, the decision of nine out of ten of those who now remain bachelors during the whole or a great part of their lives would, there can be no doubt, be in favour of marriage. If, therefore, every man among the middle and higher ranks were compelled to lead a life of stainless abstinence till he married, we may be perfectly sure that every woman in those ranks would have so many offers, such earnest and such rationally eligible ones, that no one would remain single except those to whom Nature dictated celibacy as a vocation, or those whose cold hearts, independent tempers, or indulgent selfishness made them select it as a preferable and more luxurious career. Unhappily, thousands of men find it perfectly feasible to combine all the freedom, luxury, and self-indulgence of a bachelor's career with the pleasures of female society and the enjoyments they seek for there. As long as this is so, so long, we fear, a vast proportion of the best women in the educated classes—women especially who have no dowry beyond their goodness and their beauty—will be doomed to remain involuntarily single."

Does not this suggest a work for woman to do, viz., to elevate the morals of her sex and the standard of womanhood? Let it not be forgotten that there is a reward in the future; the world knows little of its truly great ones; and there are *heroines* as well as "heroes without the laurel, and conquerors without the triumph."

But upon the third and most powerful class—the mothers—the success of the campaign here recommended chiefly depends.

> " Blessings on the hand of woman!
>   Angels guard its strength and grace,
> In the palace, cottage, hovel,
>   Oh, no matter where the place!
> All true trophies of the Ages
>   Are from Mother Love impearled,
> For the hand that rocks the cradle
>   Is the hand that rocks the world."

Heretofore, all excess of male births has been more than cancelled by male mortality, but under the old *regime*, where husband and wife so often vied with each other in their follies,—he with his numerous indulgences, and she with her frailties and frivolous fashions,—it could not be otherwise. That more males than females have resulted, only proves that man has outstripped woman in this baneful course. Let woman's aim and efforts be but in the right direction, and all wrongs can be righted; she surely holds the key of the position. Let her forsake a listless, aimless career of idleness and pleasure, and be the mistress of her home; give up one-half of her superfluous, enervating, luxurious habits and dishes, and find time to train up her children in the way they should go, with the Divine assurance that "they will not depart from it." When we consider the pernicious home influences under which so many young men are reared, it is not surprising that they become profligate. Let mothers teach their daughters to prize virtue in man, even to the extent of refusing the hand of him who does not lead a steady, worthy life.

Young women now set their hearts too much upon wealth, and demand a fortune to enable them to participate in the fashionable life they wish to lead. It is on this account that unscrupulous men so frequently adopt the course just indicated by Mr. Greg. If unmarried ladies were to insist upon *virtue* as the only possible equivalent for virtue, and if mothers would but instil the same wholesome truth into the tender minds of their children, men would ere long come to a sense of their delinquencies, and their manhood would gain the ascendancy in the course of one or two generations. Revolutions have occurred in other spheres not less unpromising, and I make no doubt that success awaits the patient endeavours of woman in this her more peculiar province.

Only a few decades are required for the regeneration of our offspring—so far, at least, as these crying social evils are concerned. But some may say : " If women elevate society in this way, and produce a large excess of sons, according to all experience this will be followed by more than a corresponding superabundance of females in the next generation." This would not affect, however, the ultimate result ; evil never results from doing right. Society is sadly out of balance because of sin and vice, but with a steady, virtuous effort, a peaceful and harmonious equilibrium of the sexes would inevitably be established. Let the plan I have inculcated be put in practice to the extent of producing the requisite number of eligible husbands, and the daughters born during the same period would be of quite as noble a stamp, for the causes indicated above as operating to produce an elevated race would come at once into activity, and influence the result.

Such, then, are some of the more remote issues which will be influenced by a knowledge of the law ruling sex in offspring; and this knowledge I have endeavoured to convey to the reader in stating my theory.

I am fully aware that the superstructure of the theory may be regarded as little more than tentative at present, but the basis on which it rests is sound; in fact, it is the only possible foundation for any theory worthy of the name, seeing that it consists of observation and fact.

The existing state of physiological science prevents my being more explicit as to the nature of " superiority," or as to the relative values of different temperaments and characteristics; the subject is in its nature somewhat complex, as, for example, the science of meteorology is complex, from the number of factors involved in the determination of the result. But the immense progress recently made in meteorology is a good augury for the future in this matter of sex, so soon as study shall be systematically applied to it, with a clue to guide its course. This clue or key, I believe, is supplied in the foregoing pages.

Dr. J. Mortimer Granville, in an article contributed to the *Lancet* of 23rd October, 1880, upon " Intention " in the determination of sex, very justly

remarks that considering the importance of the
subject, " it is scarcely creditable to science " that
so little should be known as to the laws determining sex. He also advocates a fuller and wider
research being devoted to the subject, as he believes
that " a better and fuller acquaintance with these
laws would tend to the improvement of the human
race, and the abatement of disease."

Upon the applicability of these laws when
thoroughly understood he entertains no reasonable
doubt, and then proceeds to an enumeration, under
divisional heads, of the chief fallacies of which
I have disposed in Chapter III.

In his article Dr. Granville seems somewhat
unconsciously to have approached very near to a solution of the question, but basing his arguments upon
the erroneous premiss of " predominant procreative
force," reaches a conclusion wide of the mark.

With his conclusions as to the casting of the
opposite sex by the determining parent I at once
agree, only substituting " superiority," in the sense
so often defined in this book, for his " constitutional
excess of procreative force," which, as I have often
remarked, can have no application in very many
cases. Beyond this, he so very clearly explains the
effects of inheritance in alternate generations (as
I have sought to do in the chapter on the principle
of adjustment) that I cannot refrain from quoting
him here :—

"The *seeming* absence of certain elements of the
hereditary constitution in one generation and their
re-appearance in the next or a later generation, is
not a mere accident, but precisely what we should

expect under the operation of the natural laws governing sex and heredity. Thus a father may transmit certain peculiarities to his daughter, and she will pass them on to her son, who consequently inherits from his maternal grandfather. The female link in the chain being incapacitated by her sex for the development of certain parts of the *entailed* inheritance, they lie dormant in her organism, to be transmitted, like vitalised but ungerminated seeds, to her offspring when she becomes a parent. . . . What we call 'atavism' is due to the circumstance that in the case of a male child, although *the energy of the female parent determines the sex* and does much to influence the surface type of the organism, the force of the male parent preserves the constitutional type. This enables the male characteristics to reassert themselves in any generation in which the female influence, though sufficient for the determination of sex, is not so strong as to influence the whole course of development. Therefore the latency of peculiarities of constitution is not rhythmical, but irregular, the force directly transmitted generally, however, regaining its ascendancy in every second or third generation."

He attributes considerable influence to what he designates "intention," and describes as "the purposive employment, interruption, or control of the laws known to us, for the production of a given result—such as the extinction of disease or deformity, and the engrafting of a healthy stock. It cannot be doubted that science has a wide and as yet unworked field of enterprise in this department, from which the richest and best fruits may hereafter

be gathered." Now it has been my aim throughout
to establish this point, viz., that we have the power,
by the methods indicated in the last chapter, to
elevate the race by means of precisely this "purposive control of the laws known to us," that is, by
the intelligent guidance of that "intention" of which
Dr. Granville speaks.

Indeed the whole of the foregoing chapters may be
regarded as one long insistence upon the truth contained in his concluding sentence :—

"We leave too much to chance in relation to
the propagation of the species and the *improvement
of the race.* 'Intention' has a part to play in
the determination of sex, and in the revival or repression of inherited forces, by an intelligent use
of the laws of Nature, which is not yet even
recognised."

The theory laid down in this work has been established to my own satisfaction by the most convinc-
ing of proofs—that of personal observation ; and
the especial recommendation of it as a theory is,
that every one can apply it for himself. The very
best evidence in its favour will be found by the
student among his own friends and acquaintances,
and by general observation as he moves about the
world.

At the seaside, for example, one often sees whole
families together,—father, mother, and children.
With a knowledge of the principles determining
"superiority," any one may thus daily verify my
theory for himself. He will see a portly father, dull
and lymphatic, coupled with a nervous, sprightly

wife; and with these he will find a troop of boys, inheriting, very probably, much more of the mother's temperament than the father's. On the other hand, when he meets a thin and active father, perhaps resting from the exertions of his profession, with a stout and pleasant, but indolent or lymphatic mother, he will as surely find a family of daughters who have evidently inherited the father's activity. The cases of this sort that I have met with in my travels, every one confirming my theory, are innumerable, and I doubt not that others will have a similar experience. At all events, I am content to rest my theory upon its ability, in the hands of any reasonably careful observer, to explain the facts relating to sex, with which he may meet in his daily life. By this test every theory must stand or fall, and mine differs from others only in the greater facility with which any one may bring it to this great touchstone.

I do not, however, by any means claim that my theory has been perfected, and that there is no further scope for discovery as to what determines sex; nor that I may not have committed some errors in my premisses or in my reasoning. But I do maintain, *first*, that I have conclusively shown that all previous theories are either wholly futile or based on a mere shadow of truth, of which their authors themselves have been unable to obtain a full grasp, and tha t they have therefore remained abortive or ineffectual; *secondly*, that I have discovered the true law of sex, which is the only one containing a principle of self-adjustment, or accounting for the facts of heredity which we daily see around us; *thirdly*, that my theory can be mastered by every attentive student, without special technical training, and may be confirmed unmistakably by reference to the world around him ; and *lastly*, that it confers upon the attentive student the power of predicting the sex of the offspring of any given marriage, and upon parents the power of choice of sex in their offspring in almost every case. All this it does, too, while helping to mitigate many great social evils, which it also tends to abolish entirely ; and while promoting the progressive elevation of the race, I may be permitted, in conclusion, to indulge the hope that this theory may arrest the attention of physiologists and others, and may stimulate and direct their special researches, thus contributing to a further solution of the most recondite problems, both of human and social organization.

APPENDIX.

In an article in the *Lancet* of Oct. 23rd, 1880, by Dr. J. M. Granville, already referred to in Chap. XI., the theory is put forward, but avowedly without any argumentative support, that "sex is determined by the relative ardency of the two parents." The author does not explicitly define this term "ardency," but in subsequent letters he says he does not mean "animal passion," but that it refers to energy of impulse, and, indirectly at least, to constitutional vigour; and he speaks of it as a "predominance" or "constitutional excess of the procreative force." Further he says: "A preponderance of impulse on the part of the male parent produces female offspring, while excess on the part of the female parent produces male progeny." This is a nearer approximation to my theory than any I have seen, though it is obviously different in most essential points; and coming from such an authority it will be accepted as an endorsement of the general principles of my theory. Notwithstanding the apparent approach of Dr. Granville to my theory, nevertheless, he is still wide of the mark, inasmuch as he makes intensity of feeling, or desire, as its basis; for though the definition of "ardency" is not very precise, it is obviously very different from "superiority," as I have used and defined that term. But the agreement is most striking in the application to the question of adjustment by alternate preponderance in succeeding generations, modified by the new influences brought to bear by marriage and the mingling of new blood,—extracts from which I have quoted in the text.

So far I might have mentioned this theory among those current in Chapter III., but the facts of observation adduced by Dr. Granville seemed to make some special reference necessary, which could not well be introduced in the body of the work. For the rest, I fully agree with Dr. Granville, substituting the much more comprehensive series of determining forces which I have labelled " superiority," for that which he designates " ardency." As instances of the corroboration of his theory, he adduces certain facts of observation in connection with mankind. The first, that " the children of quickly-married parents are generally females," is not very clear, and would be difficult to prove, owing to the ambiguity of the word *quickly* in this connection, and the arbitrary basis of selection for typical cases. The same may be said of the converse case also cited. Secondly he says, " Children of unions where the female is not a consenting party, are almost invariably females." This I should expect as a general rule, owing to the depressed spirits of the female parent in contemplation of a union from which she revolts, operating powerfully to lower her superiority. The author also notices that an odd child in the midst of a one-sided family is generally weakly,—a point I have long observed, and have called attention to in the text.

In a later letter (November 13th, 1880), he avows his belief in the communication of mental impressions by the mother to the child during gestation; with all of which I most fully agree.

His statements, however, on December 18th, 1880, as to the female sex being due to an arrest or repression of the force of development, or the formative, I am unable to see my way to accept, notwithstanding his declaration that this repression has nothing in common with immaturity.

The theory of Mr. Roberts, F.R.C.S., in answer to Dr. Granville (*Lancet*, December 11th, 1880), is but little different from the physical vigour theory mentioned in Chapter III., and is open to the same objections. Some of the statistics he gives are of value, and I have already commented upon them in the text.

The experiments with animals mentioned by Dr. Prieger (*Lancet*, May 7th, 1881), are precisely similar to those mentioned by M. Girou, cited in the text, and by no means necessitate the acceptance of the theory of genital vigour. The explanation I have given in connection with M. Girou's facts applies also to these.

THE END.

INDEX.

A.1.

B.1.

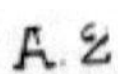

A.2.

B.2.

A 3

B 3

A 4

B 4.

A. 5.

B. 5.

A. 6.

B. 6.

A 7

B 7

A 8

B. 8

A.9.

B.9.

A.10.

B.10.

A II

B. II.

A . 12.

B. 12.

A.13.

B.13.

A.14.

B.14.

A . 15 .

B . 15 .

A . 16 .

B . 16 .

A. 17.

B. 17.

A. 18.

B. 18.

A . 19.

B . 19.

A . 20.

B . 20.